Born Expecting the Pleistocene

Mark Seely

Born Expecting the Pleistocene
Psychology and the Problem of Civilization

Born Expecting the Pleistocene: Psychology and the Problem of Civilization by Mark Seely

OldDog Books

ISBN 978-0-615-60862-4

Cover art: "Totem for Dad," mixed media sculpture, 32" h. x 18" w. x 18" d. by Bonnie Zimmer

Photograph by Juan Carlos Rodriguez

OldDog

To Skyler

CONTENTS

PROLOG

To understand everything necessary for you to understand about yourself, you need only know this one fact: you are a prehistoric forager. Your physiology, your muscles, your sensory systems, your immune system, your digestive system, your behavioral predispositions, and your emotional sensitivities are all designed specifically to accommodate the requirements of a foraging lifestyle. Your brain is wired to facilitate the navigation of a free, physically active, spiritually rich life in a small highly-egalitarian social group culturally embedded in local natural systems. This is who you are. You are a Pleistocene hunter-gatherer to the very core of your being.

The lifestyle that you have been forced to adopt is not who you are. In many ways it represents a negation of who you are. The mismatch between your physical and psychological expectations and the compulsory demands of a technology-dependent post-industrial civilization is an unbridgeable gulf. Your muscles are atrophied. Your heart languishes in a soup of stress hormone metabolites. Your lungs are brittle bags. Your eyes have lost their acuity and are blind to the periphery, and your visual world has subsequently become a narrowing tunnel. There are sounds you can no longer hear, smells that have been permanently masked, and flavors and textures that you will never know. Physically, you are little more than a walking corpse.

Psychologically, you are a child. Your goals and aspirations have been entrained to the will of an unfeeling planet-devouring machine. Your thoughts have become mechanized and outsourced. Your emotions are those of an infant.

And you are not free.

INTRODUCTION

The Modern Urban Legend

A Grim Tale

Once upon a time…

Sometime between two and five million years ago, one of the many extant species of chimpanzee developed physical characteristics sufficiently close to those possessed by the species to which modern-day scientists belong for the scientists to call them by the same first name.

This new species of chimpanzee eventually settled into a genetic configuration that is virtually identical in all important respects to yours and mine. Coincident with this configuration was a generally peaceful and fulfilling lifestyle that involved nomadic and semi-nomadic foraging in small, highly cooperative, largely egalitarian social groups, and the manufacture and use of relatively sophisticated stone, bone, wood, leather, and plant fiber tools and utensils. There were no weapons because there was no war. There was instead a lot of singing and dancing and celebration. Infectious disease was virtually nonexistent, and hunger was exceedingly rare. Many of these groups practiced small-scale gardening to augment their diet, but the majority of their food came from wild-harvested roots, fruits, nuts, berries, and game.

Time passed.
Lots and lots of time passed.

Very recently, beginning just nine or ten thousand years ago, an infinitesimally small minority of these people began to engage in large-scale domestication experiments. The incorporation of farming and animal husbandry had profoundly negative ramifications for their physical, psychological, and social well-being, as well as deadly consequences for their foraging neighbors.

With domestication came surplus food. With surplus food came an increase in population and, paradoxically, the potential for widespread famine and disease. The rapidly increasing population of farming peoples also ensured the eventual displacement or assimilation (enslavement) of surrounding foraging populations, and domestication soon became a dominant lifestyle.

Domestication was also applied to humans, yielding hierarchical power relationships and, eventually, cities and city-states ruled by kings—and lots of weapons. War and slavery and oppression and misery flourished. Early city-states dissolved and collapsed as they over-exploited their local resources. But domestication spread like a malignant cancer, and others sprang up elsewhere. And cities soon became a common feature of the landscape.

Numerous tools were developed in order to maintain and support the imposition of unnatural power hierarchies in early city-sates; chief among these tools was religion. Later tools included the application of preternatural concepts such as private property, capital, and democracy. The highly ritualized religion practiced in Europe led to the invention of the mechanical clock to coordinate the behavior of monks. Capitalists later appropriated the monks' clock to coordinate the behavior of wage-slaves, and the modern machine was born.

And the machine quickly discovered oil and the international corporation.

And the corporation declared itself the legal equivalent of a human being.

And very little time passed before the planet died.

A Machine Called Civilization

The Mesopotamian epic of Gilgamesh is one of the oldest surviving stories. It comes from a region of the world that has been dubbed the cradle of civilization, and from a time when civilization was still in its infancy. The use of the terms *cradle* and *infancy* are informative with respect to the metaphor they evoke—but I am getting ahead of myself. Consistent with the adage that all myth reflects a kernel of truth, Gilgamesh was an actual person, a king of the Sumerian city of Uruk, who lived sometime between 2700 and 2500 BCE.

Allusions to the sharp distinction and deeply antagonistic relationship between civilization and the natural world emerge at several points throughout the epic. One of the clearest examples of this antagonism is seen in the battle between Gilgamesh and Humbaba, the supernatural guardian of the vast cedar forests that in historical fact once blanketed much of the ancient Near East. Gilgamesh, the representative of civilization, decides to make a name for himself by killing Humbaba, the representative of the natural world, and chopping down the forests. Gilgamesh succeeds despite superficially overwhelming odds, and in characteristically epic form. The take-home message is that the agents of civilization are relentless and unstoppable.

A more subtle and perhaps more informative allusion to the antagonism between civilization and nature, one that also underscores the irresistibility of civilization, can be found in the story of Gilgamesh's best friend, Enkidu, a wild man living a joyous life off of the fruits of the earth until he is seduced into civilized life by a prostitute. Enkidu lives for a time with shepherds before entering the city, and befriends Gilgamesh only after Gilgamesh gets the better of him in a fight. The tale of Enkidu mirrors the story that the inhabitants of the 21st century tell about the progressive nature of their species' transition from the primitive Paleolithic past to the technologically-enlightened present (itself an interesting and informative epic legend). That Enkidu had to be seduced into leaving the comforts of nature, and that he had to be physically conquered by Gilgamesh before becoming truly civilized, serve as allegorical parallels to the transitions of actual human communities and individuals from subsistence hunting and gathering to domestic agriculture, and from farmer to factory worker, respectively.

But, once again, I am getting ahead of myself.

For now let's focus more specifically on the modern urban legend, the collection of myths that we tell ourselves about industrial civilization and our modern techno-culture. The modern legend incorporates and expands on the older myth of irresistibility, broadens the fundamental antagonism toward nature, and elevates civilization to the position of supreme beneficence. According to the modern legend, civilization is our highest achievement, the greatest of all human creations; it is the realization of our evolutionary manifest destiny, the material bloom of our superior intellect and ingenuity. And its present iteration, our high-tech global industrial society, represents the leading edge in the progressive development of the human species that began in the forests and savannas of Africa five million years ago. Civilization not only provides us with uncountable and undeniable benefits, it is the fount of everything that is grand and sublime about human nature—so much so that the words we use to describe human qualities that run counter to civilization have become invective: *primitive, vulgar, barbaric, uncivilized.* An abiding antagonism toward nature and a strong sense of separation from the natural world are fundamental features of most of our major social and legal institutions. Life outside of civilization's benevolent embrace is, as Hobbes famously proclaimed, "solitary, poor, nasty, brutish, and short," and for the technology-saturated urban inhabitants of the 21st century, literally unthinkable.

Perhaps I should clarify what I mean by *civilization.* Definitions of civilization typically refer to the presence of cities. And for most people, the word probably conjures images of cityscapes populated by tall buildings and crowded roadways. For some people, the word may invoke thoughts of music and art and sophisticated technology. But these things, cities, art, and technology, are *products* of civilization, not civilization itself. More sophisticated definitions mention the complex organization of human activity based on the division of labor, a hierarchical distribution of power, and the importation and asymmetric distribution of resources. Civilization is the application of a collection of systems for organizing human behavior; its purpose is to direct human behavior toward ends that could not be achieved in the absence of these organizational systems. A more direct way of defining civilization—the definition that I will be using throughout this book—is that civilization is a complex machine. Lewis Mumford (1966; 1970) called it a *megamachine,* a machine composed of numerous interacting mechanical subsystems, a machine whose function is the large-scale organization and manipulation of human activity in the name of some higher power or higher purpose.

I want to make clear that defining civilization as a machine is not metaphor. According to Marriam-Webster, a machine is "[a]n assemblage of parts that transmit forces, motion, and energy one to another in a predetermined manner." A machine is any device that allows for the useful modification or redirection of energy. To be a machine, it is not necessary to be composed of a particular kind of substance or to have physically moving parts. The simplest machines, e.g., an incline plane, a lever, a screw, can be composed of an infinite variety of substances and have few or no moving parts. The same is true of some of the most complex machines, the integrated circuitry of micro-computers, for example. The important thing is the organization of the parts, the functional relationships among the parts. Civilization is an actual machine; it is the broad-based application of mechanical systems to human thought and behavior in such a way that human beings, their relationships to one another and their interactions with their physical environments, serve functional roles within the mechanical systems—as literal cogs, sprockets, levers, conduits, servos, valves, converters, transistors, resistors, processors, transmitters, and relays.[1]

Civilization is a machine designed to organize, modify,

and direct human activity. Part of how the megamachine of civilization accomplishes this through the implementation of hierarchical power structures that rest on a finely partitioned division of labor and the sanctioning of authorities (or experts). The division of labor into narrowly defined slivers of expertise insures that each part, each person, remains dependent on the functioning of the structure as a whole; and the finer the divisions, the greater the dependence. The difference between the megamachine assembled in ancient Egypt to build the great pyramid in the name of the god-king and the one that is presently busy reshaping the entire planet in the name of human progress (and the almighty dollar) is a difference of scope and efficiency: modern civilization, because of factory mass production, sophisticated computer technology, complex networks of electronic communication, and intricately structured organizational systems—and let's not forget a massive amount of fossil energy—is orders of magnitude more efficient at channeling human activity. We will return to the modern version of the megamachine and explore it in some detail in Part 2.

To what ends is the machine of civilization directed? Are they human ends? If so, then whose? Which humans? They are clearly not the ends of the overwhelming majority of humans who participate as cogs and sprockets. The individual components of a machine have no choice but to act in accordance with their relationship to other components, in accordance with the design of the whole; it doesn't make sense to suggest that civilization is designed to facilitate the goals of individuals when the individuals' goals are determined by the nature of civilization itself. Nevertheless, because the system grooms us to see ourselves narrowly in terms of our functional roles within the system rather than as human beings, as economic units—as individual consumers—for example, rather than in terms of our relationships to natural human communities, this is exactly what we come to believe: civilization serves us because we serve civilization. Civilization manufactures ends for us to adopt and convinces us that these ends actually serve our own interests. And, further, at least not without large scale (overt) slavery, no civilization would be able to function unless some critical mass of the population continually acted on this belief.

Consider the following statement on the purpose of institutions, taken from chapter 2 of the National Council for the Social Studies' National Curriculum Standards:

> Institutions are the formal and informal political, economic, and social organizations that help us carry out, organize, and manage our daily affairs. Schools, religious institutions, families, government agencies, and the courts all play an integral role in our lives. They are organizational embodiments of the core social values of those who comprise them, and play a variety of important roles in socializing individuals and meeting their needs, as well as in the promotion of societal continuity, the mediation of conflict, and the consideration of public issues.[2]

The idea that the institutions of civilization, a school or a government agency, can be the "organized embodiments of the core values of those who comprise them" is a complete inversion of reality. Individuals' core values and beliefs are a product of the institutional organization of thought and behavior, the result of institutions fulfilling their "variety of important roles in socializing individuals," not the other way around. Yet, the idea that we serve ourselves by serving the machine is a deeply entrenched belief.

> In the name of individualism, civilization manufactures stereotypes: Dumb Doras, organization men, or Joe Magaracs, whose prototype, in the popular tale, is transformed into the very steel that he helps produce. Such stereotyping usually leads to a culturally formed stupidity, a stupidity of the job itself, which grows to encompass the person, feeding on itself as both a defense against experience and the result of being deprived of it. But the psychologically isolated individual, cognitively, instrumentally, and affectively dulled by the division of labor and

threatened by leisure yet somehow treasuring the idea
that, in his name, society functions and battles are
fought, is unknown in primitive society. (Diamond,
1974, p. 166)

We will be fleshing out this inversion of reality in more detail
in Part 2.

What is the ultimate purpose of civilization? Why does it
exist in the first place? In the past, these questions could easily
be answered by reference to a divine plan or to the
question-begging notion of manifest destiny, and both of these
answers still lurk prominently in the background shadows of
the modern urban legend. According to the modern legend,
civilization is both a cause and a direct result of human
progress. This of course raises the question of what is meant by
human progress. Progress implies a direction, some goal state
toward which civilization is heading, an end-in-view so that we
can gage our progression in terms of increasing proximity to
that end. What is the end to which civilization is heading?

Empirically, we might be able to get some notion of where
civilization is actually heading by taking a look at the recent
past, by comparing conditions that exist now with conditions
just a short time ago. Increased population and a concomitant
increase in hunger and general suffering, decreased species
diversity along with global deterioration of the biosphere,
perpetual war and ever-expanding arenas of armed conflict,
extensive surveillance leading to the complete elimination of
privacy, deepening alienation: these are all clear indications
that we are not headed in a good direction. But these things
aren't included as part of the modern legend, or they are
incorporated only peripherally as challenges, problems that
additional progress will eventually resolve. According to the
legend, in the past life was mired in unimaginable struggle and
inconvenience. In the past we were ignorant and helpless
against the forces of nature, especially against sickness and
disease. So, then, is the future of our legend, the end to which
our progress is leading us, a utopia in which every obstacle to
human happiness, every discomfort however trivial, has been
removed and every disease has a cure?

Or is there something other than a quest for utopia that is propelling us forward? There is another way of conceiving progress. You can progress in your pursuit of a desired goal, but you can also progress in terms of your flight away from something undesirable. Maybe civilization represents progress, not in terms of the approach to some future utopia, but in terms of increasing distance from some past torment. Is the modern legend a tale of escape? Is it a story of how we have managed to drag ourselves further and further out of the dark Hobbesian pit of primitive ignorance, fear, and suffering? The idea that our progress is in terms of an escape from primordial darkness might explain both the general disdain for things primitive and the uncritical acceptance of technological innovation that are characteristic of the modern civilized worldview. Note that both the "quest for utopia" and the "escape from the primitive" versions of the story assume that civilization accumulates benefits, that it moves us in the direction of the general improvement of the human condition, that life is getting progressively better.

Regardless of which version of progress you take, quest or escape (I suspect that both are in play), the modern legend of civilization is patently false in almost every respect. Civilization does not represent progress in any sense beyond simple accumulation. The mechanical operation of civilization is linear, perpetually expansive, and unsustainable. It imposes lifestyles that run counter to our human nature and it is permanently altering the planet in profoundly negative ways—in clear opposition to even the most generic idea of progress as the accumulation of improved conditions. Civilization has clearly made things worse. Despite this, the myth of progressive improvement has almost universal sanction. I call this feature of the modern urban legend *the progressive delusion,* the demonstrably false belief in civilization as a mode of human progress. It is becoming increasingly clear to me that this delusion, that civilization "advances," that things are getting better and better, lies at the very heart of our inability to come to grips with our most pressing environmental, economic, and social issues.

But no matter how the issue is put, at home or abroad, faith in progress as the outcome of their techniques and ideas justifies Western civilized men [sic] to themselves. One must acknowledge further, that faith is the dominant idea of Western civilization. In its name Western man rationalizes not only his self-interest, but also his failures. He does not perceive his failures as the result of his goals, nor of his motives, but rather of the means at his disposal, or of his human limitations. For he cannot surrender the notion of progress without destroying the rationale for his entire civilization. No matter how critical he may be of the realities of his society, he clings to his progressivism as he would to his sanity. It is the notion of progress that mediates his alienation, and makes it possible for him to construct a reality which he does not actually experience. (Diamond, 1974, p. 39)

In addition to the fundamental assumption of progress, the modern story of civilization assumes that civilization reflects something innate about us, that it is an emergent property of human nature, and, as such, that it is an entirely predictable outcome—that it is inevitable.

Civilization is a Black Swan

Imagine a football field in which the distance from the visiting team's goal line to the last four-fifths of an inch prior to the home team's goal line represents human existence prior to the industrial revolution. Or, to invoke a structural analogy, imagine a large multistory Victorian style house in which all of the structural space, including the foundation, the framing, the floors, the walls, the fixtures, the insulation, and the original siding, represents the evolutionary history of our species up to the onset of domestication just prior to the agricultural revolution. The remainder of our species' history would be equivalent to a layer of vinyl siding tacked over the top, with mass technology represented by a thin veneer of pigment over the top of that. The siding plays no role in the structural integrity of the house—and even less so, the paint. But four-fifths of an inch can mean the difference between a touchdown and losing the game entirely!

In his book, *The Black Swan,* Nassim Nicholas Taleb (2007) provides a compelling story about the problems of induction, the intractability of prediction, and the limits of our ability to understand the causal fabric of the universe. Taleb defines Black Swans as extremely low-probability, highly consequential events that are entirely unpredictable beforehand but easily accounted for after the fact. A nuclear meltdown caused by a tidal wave generated by a record-breaking earthquake is a paradigmatic example. Specific Black Swan events are exceedingly rare by definition, but the occurrence of Black Swans as a general phenomenon is ubiquitous. Every major event in history and every major consequential feature of industrial society is a Black Swan. Our personal lives turn out to be riddled with Black Swans as well. Civilization itself, according to this view, is a Black Swan. Civilization is an exceedingly improbable event that is nonetheless entirely explainable after the fact.

A situation known as the Fermi paradox provides a thought-provoking way of exploring the extremely low probability of civilization. According to the Fermi paradox, the probability is extremely high that there should be intelligent life elsewhere in our galaxy, which means that by now alien civilizations should have spread all over the galaxy; yet there is no sign of them (ignoring the claims of the crop circle crowd and their ilk). This apparent paradox, which is sometimes cited as proof that we are alone in the universe, is not, however, good evidence against the existence of advanced alien intelligence. If anything, it is evidence of a failure to comprehend the nature and relative time frame of our own technological civilization (and perhaps the nature of intelligence itself). In relative terms, industrial civilization represents only the last few nanoseconds of the brief hour of life on this planet. Further, because our technologically advanced civilization is completely unsustainable (i.e., founded on unsustainable environmental exploitation and the unidirectional extraction and concentration of resources—as is all civilization) it is, at best, only a short-lived blip in our species' tenure on this planet, the last visible spark of a brief smoldering flame. It may be that that is the nature of all high-tech civilization. You don't get "technological advancement" of the kind that leads to space exploration without the industrial revolution (a Black Swan). And you don't get the industrial revolution without first establishing a society that includes unsustainable environmental exploitation and volatile social conditions generated by the division of labor and the concomitant asymmetric distribution of resources, all byproducts of the agricultural revolution (another Black Swan). Also, far from being a principal indication of our species intelligence, modern civilization represents a pretty unintelligent and impoverished mode of living. And, more relevant to the Fermi paradox, if there ever were any other technologically advanced species in the galaxy, they have very likely already been and gone because of the extremely transient and volatile nature of civilization itself. Because civilization represents such a short-lived period of an intelligent species' existence, the odds that we would find other civilizations in existence during the brief window just prior to our own civilization's evaporation become exceedingly small even if technologically-advanced civilizations are fairly common occurrences throughout the universe. And if civilization is a Black Swan, the odds of it emerging on two planets at the same time and in such a fashion that they could be aware of each other's presence might be exceedingly small even given a hundred lifetimes of the universe. If civilization is a Black Swan, the lack of aliens is no paradox.

Psychologically, Black Swans lead us to overestimate our predictive abilities. Our expectations for the future are driven by prior experience. Because of this, and because Black Swans are readily "explained" after the fact, we grossly overestimate our ability to predict and prepare for the future. We are lethally overconfident. And, paradoxically, the more specific information we have about the past, the less prepared we are for what actually happens. Taleb reworks a Bertrand Russell quip into a cautionary tale: 1001 days in the life of a turkey (Russell used a chicken, and was making a somewhat different point). The story goes something like this. Suppose you are a turkey, and for the last 1000 days, your entire life to this point, the humans in your world have gone to great lengths to see that your needs are taken care of. Someone is always there to help you down when you get stuck in the apple tree. You always have enough food, your water trough is regularly cleaned, and you are given shelter from the cold and a fine strip of pasture in which to stroll during the day. When you wake up on that crisp 1001st morning, the Wednesday before Thanksgiving, you have every reason to expect more of the same. There is nothing in your past experience to prepare you for what is about to happen to you.

In the modern legend, civilization is the inevitable outcome of human history. But this is based on a backward looking explanation of our current situation. There is nothing about our species prehistory that allows us to predict the appearance and proliferation of civilization. Civilization is not a predictable outcome of any combination of human physical or psychological characteristics. It is not a predictable outcome of any combination of geological, meteorological, or geographical characteristics of the planet. In fact, had the interactions among geological, meteorological, or geographical circumstances manifest themselves in even slightly different ways than they did, civilization would have never happened. Civilization is not inevitable. Civilization is a Black Swan.

No Turning Back

According to the modern legend, civilization is part of what it means to be human. Civilization is unavoidable and unpreventable because it is an emergent property of human nature. If (when) the current iteration of civilization suffers an apocalyptic collapse, assuming that any humans remain alive, a new civilization will eventually materialize from the ruins. Civilization is who we are; it is inevitable. And even if civilization is not, strictly speaking, inevitable, now that it has occurred there is no turning back. For one thing, the human species has been physically altered by civilization. We are no longer the same physical beings that we once were. We have evolved to accommodate civilization; civilization is now our natural habitat. Researchers have found evidence of numerous genetic changes that have occurred in the last several thousand years, many of which have occurred within just the last five millennia. Our DNA has apparently been altered by civilized life. Some scientists go so far as to suggest that civilization is itself a result of critical changes in our DNA. Either way, we have evolved beyond our foraging ancestors and can no longer be expected to flourish as Paleolithic hunter-gatherers. And in a couple more generations we will have changed so much as a result of our accelerating dependence on advanced technology that even a return to low-tech Neolithic subsistence farming will be impossible.

There are numerous things wrong with the view that civilization is a direct result of evolution and with the view that we have evolved to accommodate civilization. First, these views conflate two different senses of the term evolution. Evolution in the most general sense simply means change over time. In this sense, even things such as automobile designs can be said to evolve. So can a landfill. But the word is also used to refer specifically to species-level biological changes, changes in the characteristics of a species across time that are due to geographic isolation, random mutation, and natural selection. Theoretically, culture can influence biological evolution, and vice versa, but cultural evolution and biological evolution operate by different mechanisms. Second, latent within these views is the implication that evolution is progressive, that species alive today are superior to those of past periods. This implication comes right out of the progressive delusion. Also, and perhaps more disturbing, anyone who claims that humans have evolved substantially within the last five to ten thousand years is limiting their definition of "human" to exclude the indigenous inhabitants of Australia and New Zealand and other groups of people who have until recently (in evolutionary terms) been living in isolation from the populations of Africa,

Asia, and Europe—an isolation that predates the agricultural revolution by more than 60,000 years (Rasmussen, et al., 2011). If we include Australian aborigines as members of the human species, then the changes in human DNA being cited as evidence of recent human evolution are really just changes in the DNA of subpopulations of humans, and represent local fluctuations and genetic variability between subpopulations (which occur all the time in virtually all species), not changes in humans as a species. And the changes we are talking about are really not all that substantial anyway. I suspect there are far more profound genetic differences between poodles and cocker spaniels.

And, further, I would argue that any recent changes in human DNA are a result of artificial selection, not natural selection. They are side-effects of the intentional "domestication" of the human species, alterations in gene frequencies resulting from millennia of war and genocide. And as for whether our ever-changing DNA prevents us from flourishing in the absence of civilization, it is perhaps most informative to note the speed and seamless ease with which domestic dogs, creatures with at least 32,000 years' worth of intentional genetic manipulation, revert to their pre-domestication pack-scavenging ways when given the slightest opportunity. The inevitability myth doesn't stand up to biological fact.

The idea that civilization can't be undone also has a hard time standing up against actual historical fact. There are numerous cases of civilizations that dissolved or were abandoned, with the people reverting to life-ways that resembled what they were doing prior to civilization's onset. The Mayan civilization is a paradigmatic example. The history of the Mississippian culture in North America is particularly informative in this regard as well. Domestication of corn led to an increase in the local population. Ancient Egypt in microcosm, the extra people were put to work hauling baskets of dirt to erect large pyramidal mounds atop which the priest-kings would command that even more dirt be hauled and even more land be cultivated. Eventually, probably partly as a side-effect of the poor nutrition that results from a corn-based diet, the people abandoned the mound cities and went back to the more reasonable communal hunting and fishing lifestyle that their neighbors never left off. All that remains are the piles of dirt and their story written in archaeological artifacts.

An additional facet to the inevitability myth is the apparent lack of any truly viable alternative. The global environment has changed in ways that seem to preclude the wide-spread practice of subsistence lifestyles. Industrial toxification of land and fresh water is so pervasive that it would be impossible for large numbers of people to adopt non-civilized life-ways. Returning to a non-civilized form of existence would require too much sacrifice as well. For one thing, if the industrial infrastructure were suddenly withdrawn, billions of people would die almost immediately. The planet cannot support seven billion people living subsistence lifestyles. We long ago exceeded the carrying capacity of the natural environment. We are left with no choice but to continue with the status quo.

Interestingly, alternatives to the status quo continue to exist. Glaring evidence that civilization is not inevitable is found in the continued presence (albeit in vanishingly small numbers) of non-civilized peoples. If civilization is inevitable, then how do we account for the Awá people of Brazil or the Hadzabe of Tanzania? And how can we explain the directionality problem associated with historical points of contact between "civilized" and indigenous cultures? As Rousseau and others have noted, history is replete with examples of people abandoning civilization to live with the natives but few if any stories of the opposite. Westerners frequently "go savage" but the indigenous never freely choose civilization. Civilization has always had to be forced on the "primitives." Non-civilized peoples do not willingly give up their subsistence-living ways. The conversion to civilization has always involved coercion, violence, and, in many cases, outright genocide. People have to be forced into civilized life because it doesn't come naturally. This last feature of civilization, that it is not voluntarily adopted, that it is invariably imposed through coercion, compulsion, and lethal force may provide some insight into the true source of the inevitability myth: for those on the receiving end of an overwhelming power differential, resistance means oblivion.

> Civilization originates in conquest abroad and repression at home. Each is an aspect of the other. Anthropologists who use, or misuse, words such as "acculturation" beg this basic question. For the major mode of acculturation, the direct shaping of one culture by another through which civilization develops, has been conquest. Observe, for example, the Euro-American "influence" on the South Vietnamese, or the fact that white Protestant Anglo-Saxons who settled in New England learned a few things from the aboriginal Indian peoples (of which the writers of history for civilized children make so much), and the black slaves "contributed" African rhythms to southern American music. [...] This historical fact [of conquest by the invading civilized culture] is then reflected as a law of development; as civilization accelerates, its proponents project their historical present as the progressive destiny of the entire human race. (Diamond, 1974, p. 1)

Civilization is not inevitable. Yet the meme of inevitability is repeated constantly in the media. It is especially salient in discussions of technology and technological "advancement" (what, I wonder, is the ability to evaporate Pakistani children with a robotic aircraft piloted by someone drinking Mountain Dew in an air conditioned trailer in Colorado an advancement toward?). It is also seen in the laissez faire approach to corporate exploitation, complete acquiescence to corporate privilege, and the implicit assumption woven throughout our legal code that "needs" of corporate entities supersede the (real) needs of individual persons. The media propagate a malignant acceptance of the deterioration of the natural world even as they gleefully report on the latest environmental catastrophe, species extinction, climate change estimate, or industrial toxin. It's just the price of progress, after all. It's the progressive delusion again. "Progress is inevitable. Can't you see? It's right there in the word progress!"

The presence of civilization is an unlikely event, a Black Swan. Why it occurred, and why it occurred when it did are among the greatest mysteries of the human situation. Civilization accounts for less than one percent of the timeline of human existence; all civilizations in the past have proved transient; there are presently existing alternatives; the people practicing these alternatives are living happy and fulfilling lives; civilization is maintained only through violence and the threat of violence; and it has to be violently foisted onto the non-civilized. The inescapable conclusion is that, as a principle of social organization, civilization is not one of our species' defining features. It is a quirk, an accident of geography or climate or viral invention, or some unlikely combination of circumstance. It is important that we divest ourselves of the teleological illusion that civilization is a natural outcome of our evolution, because until we do, we will continue to accept civilization and all of its concomitant suffering as natural and unavoidable, as part of the price we pay for being what we are. The existence of civilization cannot be used as evidence of its inevitability. Nor is it evidence of its intrinsic value; nor is it an argument for its continued existence.

In the big picture, however, whether civilization is inevitable is really a secondary consideration. The fact of the matter is civilization is a huge problem—and not just for us humans, civilization is a huge problem for all life on the planet. Civilization is a global death sentence. Drastic action is needed very, very soon.

Technology to the Rescue

The omnipresence of technology is the one feature of the modern megamachine that most clearly distinguishes it from earlier versions. We live in a technologically-saturated and technologically-dependent society, a *technopoly,* according to Neil Postman (1992). The industrial revolution marked a critical transition in the operational design of the megamachine in which human power became increasingly replaced by mechanical surrogates, and humans were increasingly demoted to the peripheral roles of maintenance and oversight. The computer revolution has since largely displaced these latter roles as well, and humans are becoming increasingly superfluous, and in many cases, costly impediments to the machine's smooth operation. The eventual elimination of any meaningful role for humans in the megamachine's physical operation is on the near horizon.

This will not be a problem for the machine. The prime function of the megamachine is the organization and control of human behavior. What better way to exercise this control than to relegate humans to the role of end user? It is important to keep in mind that not all behavior involves overt physical activity; mental activity, cognition and emotional responses, are modes of behavior as well. Likewise, not all technology involves concrete, physical machines. Abstract legal, economic, and social institutions are technologies. Corporations are technologies. Governments are technologies. Pedagogical systems are technologies. International trade agreements, fast food franchises, and compound interest rates are technologies. I have much more to say about technology, and about our relationship with technology, in Part 2. For now, I will set the stage by briefly addressing another essential feature of the modern urban legend: two pervasive myths about the nature of modern technology, the myth of technological progress and the myth of ultimate beneficence.

Technology lies at the heart of the progressive delusion. The progressive nature of technology appears to be a demonstrable fact. Just look at the increasingly comprehensive ways in which technology is being applied in our lives. Technological progress occurs through innovation. Innovation is an unassailable good. The term *Luddite* is an insult leveled against anyone who questions the value of technological innovation, anyone who demonstrates reluctance toward new technology, anyone who fails to ascribe whole-heartedly to the progressive delusion. The historical Luddites were early nineteenth century textile workers in Britain who organized against the increasing automation of the textile mills. Their actions resulted in legislation making the destruction of a machine an offense punishable by death, and several people were hanged for the sole crime of damaging mill machinery. According to the megamachine's logic, the structural integrity of a mechanical device is more valuable than human life. Woe to those who try to block the path of technological progress.

The progressive nature of technology appears to be a fundamental feature of technology itself. Technology is cumulative; older technologies provide the foundation for new and improved versions, and prepare the way for additional innovation. Once an innovation occurs, once a new technological application is created, the ratchet sets, and the new technology becomes a permanent part of an ongoing and ever-expanding process. The problem here is that *accretion* and *accumulation* are not the same as *progress*. As discussed earlier, progress entails a target or goal, an end to which current conditions can be compared. What is the end goal that allows us to gage technology's progress? What is the unit of measure? Changes in technology are frequently referred to as "advancements," and it is true that, within a strictly mechanical-production frame, new technology is frequently an improvement over old technology (i.e., it is faster, more efficient, etc.). But this interpretation of advancement does not translate to anything meaningful outside of a limited mechanical-efficiency frame.   What are we advancing toward? What are we advancing from? How is an increase in the capacity to consume resources, curtail individual freedom and privacy, increase population, wage war, alienate ourselves from the natural world, and impoverish the biosphere an improvement over previous technological conditions? If there is any sense in which technology is progressive, outside of a strictly mechanical-efficiency frame, it is only as a synonym for accumulation. In the accumulation sense, technology is indeed progressive, but progressive in a decidedly undesirable way: it is progressively corrosive of the human condition as a whole.

The progressive nature of technology is a feature of the internal logic of the machine, with its narrow focus on speed and efficiency. As with civilization more generally, technology is, in Mumford's words, "absolutely irresistible." It is an emergent property of human intellect. It is an unavoidable feature of humankind's cultural development. This myth, like its parent, the myth of the inevitability of civilization, is extremely difficult to dispel despite the fact that it has very little in the way of either logical or empirical support. Logically, the Black Swan problem applies: the belief in the inevitability of technology is supportable logically only through inductive reasoning specifically based on a history-textbook view of Western civilization that ignores numerous actual cases in which technological innovations were never implemented or were abandoned in favor of more primitive versions, and numerous whole-scale societal reversions to lower-tech life-ways.

The second pervasive myth about technology that I want to touch on here is the myth that technology is ultimately beneficial. Even the most hardcore technophile has to admit that technology can create intractable problems (nuclear waste, anyone?). Despite this, technology is seen as an ultimate good. This myth is closely tied to the myth of progress. Who could deny that technology is making life progressively better? The problem with this is that "better" is not a stand-alone generic term. Things are never generally better; they are better in specific ways. And by restricting the conversation to the specific ways that a limited set of conditions is better, the myriad ways in which things may have become worse are often completely ignored. Once again, this is an example of confusing accumulation with improvement.

> No Society can escape the fact of change or evade the duty of selective accumulation. Unfortunately change and accumulation work in both directions: energies may be dissipated, institutions may decay, and societies may pile up evils and burdens as well as goods and benefits. To assume that a later point in development necessarily brings a higher kind of society is merely to confuse a neutral quality of complexity or maturity with improvement. To assume that a later point in time necessarily carries a greater accumulation of values is to forget the recurrent facts

of barbarism and degradation. (Mumford, 1934 p. 184)

Cost-benefit analyses of specific technologies frequently ignore costs that are difficult to quantify, costs that lie outside of the immediate purpose, proximate processes, or time-frame in which the technology is being employed. These peripheral costs are called externalities by economists because they are costs that are not born directly by the owners or users of the technology. Environmental pollution and the general degradation of the quality of life on the planet are distributed costs that are difficult to factor into an accounting ledger. The failure to account for externalities provides us with a biased view of a technology's effects, a bias loaded heavily toward the positive side of the ledger.

This positive bias is particularly conspicuous with medical technology. Where would we be without polio vaccine or Prozac or dialysis? Clearly medical technology has a demonstrably positive impact on the quality of life. Let's ignore for the moment the millions of individuals with medical problems that require technological intervention who will die because they do not have access to the technology, either because they are too poor or because they had the misfortune of being born on the wrong part of the planet—or both. The positive bias leads us to downplay or ignore entirely the real damage medical technology causes. We call specific instances of this damage *side-effects* as if they are not a meaningful result of the technology's application, and thus not to be given full weight in a cost-benefit analysis. In addition the number of people (with access) whose lives existing technology will be able to extend or improve does not begin to offset the suffering of people currently inflicted with medical conditions caused directly or indirectly by life in a physically and psychologically toxic industrial society. When you consider that the overwhelming majority of medical conditions that require treatment using advanced medical technology are themselves direct or indirect results of our dependence on technology, the argument that medical technology is making things generally better dissolves entirely. Consider, as salient examples, diabetes, heart disease, depression, and most types of cancer. For years I have been intrigued by the absurdity of exposing people to the carcinogenic effects of X-rays as a way of detecting lung cancer. Before anyone suggests that modern technology does more good than harm, they need to first weigh the costs and benefits associated with advanced medical technology—taking care to include the hidden physical and mental health costs associated with the corporate industrial infrastructure that serves as a precondition for modern medical technology's existence in the first place. You don't get MRIs or antibiotics without a toxic environment and a crowded, stress-filled, nutritionally-deficient modern lifestyle. The need for advanced medical technology is a direct byproduct of the conditions that support its very existence. And the increase in need has always outpaced medical tech's ability to keep up. How much of an impact is modern medical tech going to make in Japan, as tens of thousands of people start to manifest mid- and long-term symptoms of radiation poisoning resulting from the Fukushima nuclear meltdowns? Whatever the impact of medical tech turns out to be in this specific case, the resulting health of the affected population will be far inferior to what it would have been had the nuclear power plants never been built.

Contrary to the myth of beneficence, technology is a zero sum game. What's even worse is that technological innovations never actually solve any of the problems they were designed to deal with. Technology never solves anything. It never has. It doesn't solve problems so much as it redistributes them. The net result is null. The zero-sum nature of technology is a variation on the first law of thermodynamics (conservation of energy), which holds that, in an isolated system, energy can neither be created nor destroyed. When it comes to a specific problem that a given technological innovation is supposed to solve, it is possible to change the problem's form, or redistribute the problem in time or space, but the total amount of "problem-ness" remains, and will eventually turn up somewhere else, pleading the development of yet additional technological innovations.

As a concrete example, consider the problem of "superweeds" that have evolved resistance to the herbicide used on corporate corn and soybean fields in the Midwest. The farmers (really outdoor factory workers) are now left with the choice of using an even more toxic chemical, plowing more frequently, or (gasp!) pulling the weeds by hand. One farmer was quoted as saying that technology-wise this development has knocked them back 20 years (Neuman & Pollack, 2010). That's not exactly true, though. Twenty years ago these superweeds didn't exist.

The specific herbicide (Roundup) is a technological innovation designed to address a specific problem. But its use didn't solve the problem. It merely redistributed the problem, changed its form, and pushed it into the future. The original problem was pesky weeds. The herbicide killed the pesky weeds (one problem), but increased the amount of toxins in the environment (another problem), and in the process, set the stage for the emergence of these superweeds (the original problem returns in a more "problematic" form). Technology can redistribute a problem to other places and to other people, but the overall quantity of "problem-ness" is always conserved.

That's how all technology is applied by the modern corporate machine: as a way of generating a short-term benefit for some limited group of people by pushing problems off onto others. The owner of a mining company becomes wealthy through the extraction and sale of mineral resources. It is impossible to extract mineral resources without damaging or permanently destroying—impoverishing—the surrounding environment. It is impossible to extract mineral resources without generating waste and contributing to global climate change. These "costs" are externalities that do not factor into the company's fiscal calculus. They are burdens shouldered by the general public and by future generations. And the fiscal scales are loaded heavily in favor of the machine: less tangible costs associated with human labor are given little or no weight. A miner trades his single most precious personal resource, the minutes and hours of his very life, for wages. The inequity of this trade is incalculable. When you include the dehumanizing impact of wage-labor along with the material externalities, the zero sum nature of the wealth-impoverishment process emerges as a mathematical imperative.

There is a historical sense in which our technology can be said to have led to general improvement of conditions. We can do things now that we couldn't do in years past. We can cure diseases that used to wipe out whole populations. We can communicate and travel with speed and efficiency that would make our great-grandfathers dizzy and confused. We can peer into the outermost regions of the heavens and the innermost regions of the substance of matter itself. Through accumulative changes in our technology, we have acquired abilities and capacities that we did not have previously. But outside of this historical, backward-looking perspective, can it truly be said that we have progressed? Was the telegraph achieved in order to invent the telephone? Was radio devised on the way to creating television? Was the adding machine created with the personal computer as an end in view?   At the risk of beating a dead horse, let me repeat that progress requires a goal. I can progress on my way to the top of a mountain. I can see that I am making some progress in paying-off my mortgage. But to where is our society's technology progressing? What is the end-in-view such that we can look at where we are today, compare it with where we were yesterday, and say that we have made progress toward achieving this end?

A group of French psychologists (Osiurak, Jarry, & Le Gall, 2010) recently offered an interesting theory about the nature and direction of technological progress. Technological innovation, according to the theory, is directed primarily at reducing the need for body action. Humans have the capacity to view bodily movement as a kind of problem to be solved: how can I get from point A to point B with the least amount of physical movement?   The problem is that any technological innovation that reduces physical movement is still going to require some degree of body action, a remote control, for example, still requires you to push a button, and this residual degree of body action resets the bar for additional innovation aimed at reducing body action further. I'm not sure that I entirely agree with their major premise about the general problematic nature of body action. There are too many examples of situations in which humans seek to intentionally increase body action (e.g., dancing, exercise, sex). But the idea that the reduction of body action is a target for technological innovation is intriguing nonetheless. And it answers the question about where technological innovation is ultimately taking us in a way that is both compelling and disturbing.

Maybe what we mistakenly call progress is just a drive toward ever-increasing complexity. Will the increase proceed without end? Will there be some point of no return at which the complexity exceeds our ability to control our own technology? Have we already passed that point? Is the ultimate goal a state in which we have been completely eclipsed, replaced by our technology? Perhaps it is not a drive toward complexity that lurks behind technological change, but, instead, a particularly insidious manifestation of entropy. Perhaps the continual increase in the complexity of our technology is indicative of our slow disintegration into chaos. Or maybe we can rescue the idea of technological progress by focusing on the progressive increase in knowledge that technological innovation both represents and facilitates. Some have estimated that the sum total of human knowledge doubles in less than a decade (Winner, 1977). Surely this increase in knowledge is a kind of progress, right? Not necessarily. It is clearly beyond the capacity of any one individual to have contact with even a tiny fraction of the knowledge available. And this tiny fraction is reduced in proportion to the increase in knowledge. The exponential increase in total knowledge implies a concomitant increase in personal ignorance. An increase in complexity implies a proportional decrease in our individual participation.

The idea of progress is not limited to technology or to civilization in general. It taints our thoughts about other aspects of modern society as well. Consider the grossly over-used two-word sound bite, *economic growth.* What does it mean to say that the economy is growing? Unlike the idea of technological progress, where there is some actual change, in the direction of increased complexity if nothing else, economic growth appears to be entirely a matter of accumulation; a growing economy means more business, more consumption, more jobs, more money being spent, more money being made, more people getting rich. There is no end-in-view. There is only the incessant drive toward accumulating more and more and more and more. A few wise souls have come to realize that economic growth has nothing to do with progress, and further, that it cannot continue at an exponential rate indefinitely. These prophets of doom plead for a sustainable equilibrium. So far they have failed to convince anyone. They are fighting an uphill battle against the powerful delusion of progress.

Human civilization is not heading for some ultimate culmination of history. Civilized society has not progressed anywhere since the first hominid shared food with his neighbor. It has changed. But this change is not change *toward* something. It is not change in a specific predetermined direction. It is change that has accrued though an accumulation of the effects of circumstances—political, geographical, and meteorological circumstances. Progress is a mirage that keeps us staggering around in circles like a person delirious with thirst chasing the shimmering hope of an oasis that continually evaporates into the sand. The modern tale of civilization is a myth born of ignorance, delusion, hubris and hindsight bias. Civilization is a serious problem for us. It is our *most* serious problem. It is in fact the worst thing that has ever happened. Every ostensive benefit that civilization provides is off-set by countervailing and benefit-annihilating costs that yield a net deficit in the quality of human life and irreparable damage to the natural world. There are no long term solutions to the problems caused by civilization—civilization is itself the problem. Civilization is not part of a divine plan. It is not our manifest destiny. It is not the inevitable result of human intelligence. Nor is it a natural product of human social evolution. And, what is most important—the driving motive behind this book—civilization is deeply discordant with our human nature; a life embedded within the suffocating walls of civilization is a life that disfigures, minimizes, or rejects everything that is truly human about us.

Not our Natural Habitat

The Mismatch Hypothesis

Our bodies including our brains—and thus our behavioral predispositions—have evolved in response to very specific environmental and social conditions. Many of those environmental and social conditions no longer exist for most of us. Our physiology and our psychology, all of our instincts and in-born social tendencies, are based on life in small semi-nomadic tribal groups of rarely more than 50 people. There is a dramatic mismatch between life in a crowded, frenetic, technology-based global civilization and the kind of life our biology and our psychology expects (Ornstein, & Ehrlich, 1989).

And we suffer serious negative consequences of this mismatch. A clear example can be seen in the obesity epidemic that has swept through developed nations in recent decades: our bodies evolved to meet energy demands in circumstances where the presence of food was less predictable and periods of abundance more variable. Because of this, we have a preference for calorie-dense food, we have a tendency to eat far more than we need, and our bodies are quick to hoard extra calories in the form of body fat. This approach works quite well during a Pleistocene ice age, but it is maladaptive in our present food-saturated society—and so we have an obesity epidemic because of the mismatch between the current situation and our evolution-derived behavioral propensities with respect to food. Studies on Australian aborigines conducted in the 1980s, evaluating the health effects of the transition from traditional hunter-gatherer lifestyle to urban living, found clear evidence of the health advantages associated with a lifestyle consistent with our biological design (O'Dea, Spargo, & Akerman, 1980; O'Dea, White, & Sinclair, 1988). More recent research on the increasingly popular Paleo-diet (e.g., Frassetto, Schloetter, Mietus-Snyder, Morris, & Sebastian, 2009) has since confirmed wide-ranging health benefits associated with selecting food from a pre-agriculture menu, including cancer resistance, reduction in the prevalence of autoimmune disease, and improved mental health.

There is an additional dark side to the mismatch. Many behavioral tendencies that were quite useful in our evolutionary past may actually be a direct threat to our future survival as a species. I will have much more to say about this as we continue. Future threat aside, humans (and all creatures) are better off living in ways that are consistent with their evolved capacities and predilections. We are "hardwired" to accommodate lifestyles and community conditions quite unlike those imposed by the machine of modern civilization, and our lives are impoverished as a result. Civilization is inconsistent with our human nature in at least three ways. First, industrial civilization enforces lifestyles that are increasingly disengaged from the natural world. Second, it engenders the formation of unnatural and diminished social relationships. And third, it compels us to adopt artificial, nonhuman goals.

Our psychology, like our physiology, has evolved for a lifestyle embedded in nature, a nature with which the average hostage of the Western world has only very indirect contact. Psychotherapist Chellis Glendinning (1994) claims that we are all suffering from the multifarious effects of post-traumatic stress generated by the large disconnect between our genetic preparation and the requirements of life in industrial society. In recent decades, some psychologists have begun to call themselves ecopsychologists and incorporate the natural world into their treatment of mental disorders, an approach to psychotherapy known as ecotherapy. Ecotherapy is based on the idea that many if not most of our modern mental health problems result at least in part from a detachment (estrangement, alienation) from the natural world. The only route to a permanent cure is to somehow reintegrate ourselves with nature. The problem with the ecopsychologists' program is that the natural world no longer exists in anything resembling the natural world our DNA expects: the land, the waterways, the air, the food we eat, the few animals we still have contact with, the way we conceptualize time and partition our days and years have all been dramatically altered. So we are doomed to live with the symptoms of posttraumatic stress, symptoms that are bound to intensify with each successive generation as the natural world continues to recede from our awareness.

Richard Louv, in his 2006 book, *Last Child in the Woods,* outlines several features of modern culture that limit the nature-exposure of contemporary children, both in terms of quantity and quality, severely curtailing the opportunities kids have to interact with nature in meaningful ways. With distractions such as, television, the internet, video games, cell phones, movies, and between programmed activities such as school, organized sports, and music lessons, kids simply don't have much time or incentive to play outside. And when they do have time, the privatization of natural places and parents' media-enhanced fear of child predators ensure that their play remains isolated within the confines of a fenced-in well-manicured (frequently toxic) back yard. The result is a lack of awareness and understanding of natural processes and a lack of sensitivity to the natural world. Louv reports one study, for example, that found the average 8-year-old was better able to identify Pokemon characters than the most common native species of plant and animal in the community in which he or she lives. We are raising a generation of nature illiterates, people who have no sense of connection with nature and very little capacity for empathy toward the natural world, people who suffer psychological symptoms of what Louv calls *nature deficit disorder.*

The changing role of place in our lives serves as additional source of alienation from nature. Place is the ultimate source of culture. The local environment—the local climate, geography, flora, fauna—sets the parameters regarding the specific behaviors required for survival. Culture is merely(!) the organization, elaboration, and stylizing of these behaviors over time, and the assimilation, adaptation, and sharing of these behaviors by people emigrating from different places. Obviously this is a gross oversimplification. My point is simply to indicate the foundational role local environments play in traditional human culture. Traditional culture is grounded in the physical world, in the demands and opportunities associated with specific physical places.

But all that is long past. Our connection to place is rapidly dissolving. Place is now just a synonym for location. "Location given only by grid coordinates, as in the names of nearest street junctions, would be intolerable to hunter-foragers" (Shepard, 1982, p.106), and given in numerical latitude-longitude GPS code, would be entirely incoherent. Modern "culture" forces us into lifestyles that are increasingly untethered from the local physical world. My food comes from 1500 miles away. My children are forced to move to other time zones to find employment. Events in China have a direct impact on my economic well-being. The quality of the air I breathe is controlled by lawyers and lobbyists for corporations headquartered in the Cayman Islands. As citizens of a culturally-diluted global society, we are losing our sense of place. We are rapidly becoming a race of beings who literally don't know where we belong in the world.

We are also becoming a race of beings who don't know each other. The machine of civilization grooms unnatural and mechanical interpersonal social relationships. The sheer number of people that we share space with grates against our evolutionary preparation and encourages superficial associations and shallow attachments. Research puts the number of people we can reasonably incorporate as a meaningful part of our personal lives at about 150 (e.g., Dunbar, 1993; Hill & Dunbar, 2003). There are individual differences, to be sure, but the limits in our ability to process and retain information that are imposed by the size and complexity of our cerebral cortex prevent us from knowing personally (beyond just a name and a few isolated facts) more than around 150 other persons—about the number of people in a large, well-established indigenous tribe. But we haven't been strictly limited by the size and complexity of our cortex since the advent of written language. And now, with our internet-based personal networking gadgets, we can manage the names, faces, and continuously updated trivial life details of hundreds, even thousands, of "friends." It's an obvious quantity-for-quality trade-off reflective of our general mass-consumption approach. We once lived in close contact with people who directly supported our physical existence and provided the raw material out of which we constructed life's meanings. Now we live in giant tribes of two-dimensional beings, engaged in a shared superficial monologue, searching for constant distraction, desperately trying to convince ourselves—through sheer quantity of experience—that our shallow consumption-driven lives are meaningful.

We engage each other not as human beings but in terms of the roles we play in the machine. We *network*. We *connect*. We establish *contacts*. We process each other. We are commodities to each other. Consider the stereotypical customer service experience. Consider the scripted greetings and small-talk between coworkers who pass in the hallway. Consider also how parent-child relationships are mediated by state-mandated education. Consider how marriage has become as much an economic as it is a social relationship—again, one that is mediated by the state and whose induction and, all-too frequently, resolution, require the official sanction of "authorities."

Oh yes, authority. Authority is so much a part of our daily experience that it difficult to see that it is an artificial construct, one that emerges directly from the hierarchical structuring of the megamachine. Archaeologists and anthropologists have amassed considerable evidence that, allowing for large variability among different specific groups, life in "primitive" band society is a highly egalitarian affair. Leaders get their position as a function of age and experience. But leaders in a band society are not rulers. They may be called "chiefs" but they are not bosses in the modern sense. They usually have no means of compelling others to follow their lead or even to listen to their counsel. And their leadership is frequently situation-limited and contextually-bound. There might be one leader during a fishing expedition, and another when tracking caribou, and yet another when harvesting fruit from a specific kind of tree. There are seldom hard and fast gender roles either. There are biologically-supported divisions of labor, to be sure, but even these are normative expectations rather than inviolable rules.

In sharp contrast to modern civilization, the adult members

of band society are all expected to act like grownups; and they are treated accordingly. One of the primary methods modern civilization uses to maintain obsequiousness to the rule of authority, a method that has the additional effect of stimulating and maintaining high levels of compulsive consumption of nonessential goods, is by infantilizing the population: emotional immaturity, child-like dependence, and childish impulsivity are encouraged and expected of all adults.

> Infantilization acts to reinforce the preference for the private and the puerile by treating the impetuous, grasping child as the ideal shopper, and the shopper as the ideal citizen. It inculcates in adults an obligation to give free rein to the "I want!" and "Gimme that!" that both disclose and constitute the infantile id. More than simply an option, puerility is regarded as a necessity of capitalism's survival and hence a mandate of the zeitgeist—which is, of course, the ethos of infantilization. (Barber, p. 134)

In addition to being an ideal shopper, infantilized adults are not likely to question the rules, and readily assume a subordinate posture when admonished or reprimanded for failing to act appropriately.

One of the most insidious ways that our interpersonal relationships are impoverished is through mediated forms of communication. With social networking sites and cell phones and email and Skype, technologically-mediated communication has become the norm. Our psychology is designed for face-to-face communication, and any form of mediation reduces both the informational and the affective content of the message. Face-to-face communication is information-rich, consisting of not only the surface linguistic meanings of what is spoken (the propositional content of message), but also of information from voice inflection, facial expression, hand gesture, eye gaze, body language, other body cues (e.g., general appearance, ethnicity), shared physical context (place), shared temporal focus (present). In addition, face-to-face communication includes the potential for real-time clarification and confirmation of the validity of the source (you know with whom you are communicating). Contrast the informational content of co-present face-to-face communication with a telephone conversation that provides only surface linguistic meanings with potential for real-time clarification, voice inflection (limited by signal strength and resolution), and shared temporal focus. Electronic text (text message/email/ Twitter) represents an even more impoverished mode, providing only surface linguistic meanings with delayed clarification. For the current generation of young adults, internet social networking sites serve as a primary vehicle for peer communication. The various "features" of a site such as Facebook, for instance, not only restrict the potential richness of the communicative act through mediation, they actually create entirely novel mechanically-structured communicative modes, shaping the message to fit the machine rather than the other way around. And, given the avatar quality of social network profiles, and the game-like layout of the medium itself, I'm not sure that these sites really involve communication in the traditional sense of real people sharing information with each other.

Mediated electronic communication allows us to treat each other as mechanical objects, which fits nicely with the goals of the megamachine, but flies in the face of our evolutionary heritage as active participants in close-knit human communities. Also, there is a homogenization and simplification, a dumbing-down of experience that happens as we are forced to interact more and more with an electronically-presented world rather than the world of actual experience.

Those seeking the sanctuary of electronic screens and headphones may imagine themselves seeking out the diversity of what is offered in games, films, music, and eluding the sameness of the outside world; yet bought electronic content is far more homogeneous and limiting than the actual pluralism of our natural life worlds, even if for some people it also feels more vivid and "real." At their best, movies cannot be more heterogeneous and varied than the real worlds they aspire to capture. All of Hollywood at its best is not equal in variety or originality of a single summer day's walk in a public park. (Barber, p. 225)

It has been suggested that we have become "tools for our own technology," and that our daily activities and moment-by-moment behavior are tailored to fit the demands of our technology rather than the other way around. Nowhere does this seem more apparent than with our communication technology. We are glued to our smartphones and other electronic playthings. We spend hours interacting with words and pictures on a flat screen encircled and interrupted by commercial advertisements, and convince ourselves that we are having a meaningful social experience. It bears repeating that adopting the role of end user increases our dependency and enhances the machine's control.

The third way that the requirements of civilization are inconsistent with our evolved predilections is that civilization forces us to adopt artificial, nonhuman goals, and to pursue ends that are not really in our own interests. If you catch a wild animal and put it in a cage, it is not uncommon for the animal to start acting in a way that seems entirely out of context. A captured squirrel might start building a nest, for instance. Psychologists refer to this as *displacement activity.* The squirrel's natural inclination is to escape its imprisonment and run away—that is its number one behavior of choice given the context. But it is unable to satisfy its natural inclination, and rather than do nothing and stew in its nervous juices, it runs down the list in its behavioral repertoire until it finds an activity that it can perform within the limited confines of its immediate situation, and does that instead of what it really wants to do. Caged wild animals are not the only creatures who engage in displacement activity. Humans living within the confines of civilization frequently encounter situations in which social convention or the actual physical environment prevents them from performing their behavior of choice, and so they select another activity to do in its stead. Eating is a common displacement activity. I am upset at my boss, but rather than scream at him and risk losing my job, I go to the fridge and grab a bowl of ice cream, or grab a beer and turn on the television, or buy something online, or update my Facebook status, or....

There is a higher-order class of displacement behavior that some have referred to as *surrogate activity* (see Kaczynski, 2010—yes, that's right, Ted Kaczynski, the infamous Unibomber—for a cogent and disturbing discussion of surrogate activity). Civilization prevents us from pursuing many of the goals that would be natural for us to pursue if we were hunter-gatherers, goals that are tied directly to our immediate community and our relationships with others embedded in the rich social fabric of shared meanings, goals that would be entirely consistent with our behavioral predilections. And so we adopt surrogate goals and pursuits, goals that leave us with the impression that we are doing something meaningful but are never entirely fulfilling, goals offered up by the machine. These can include anything from climbing the corporate ladder to writing poetry to developing nanotechnology—in fact, virtually all activity in industrial society that is not directly related to biological necessity (and even much of that) is surrogate activity, activity that we engage in because of the design of the system rather than the design of our own beings.

To summarize, there is a dramatic mismatch between the lifestyles to which we are predisposed by our evolutionary heritage and the lifestyles imposed on us by modern industrial civilization. Life in modern civilization leaves us alienated from the natural world, detached from each other, and distracted by the pursuit of superficial goals. But it doesn't have to be this way. Modern industrial civilization is not inevitable or unavoidable. It is not a requisite feature of human nature. Nor is it a progressive improvement over non-civilized modes of existence. And, by the way, it is rapidly destroying all life on the planet. There is really only one conclusion to draw from this: we need to do something else.

But what is it we need to do? How do we stop doing what we are doing now? Where might we look for potential answers?

The Psychological Frame

Asking the Right Questions

The way a problem is framed determines the nature of the solutions that emerge, which is just another way of saying that the answer depends on how you ask the question. Consider the following experiment (adapted from McNeil, Pauker, Sox, & Tversky, 1982). Participants were given a scenario describing a disease outbreak that was expected to kill 600 people, and then asked to determine which intervention program should be implemented. In one condition participants were told that if program A was adopted, 200 people would be saved, and if program B was adopted there was a one-third probability that all 600 people would be saved and a two-thirds probability that no one would be saved. In this condition 72% of the participants preferred program A. In a second condition, a different group of participants were asked to choose between program C, in which 400 people would die, and program D in which there was a one-third probability that nobody would die and a two-thirds probability that all 600 would die. In this condition, only 22% preferred program C, which has an outcome identical to program A. By framing the question in terms of how many people would die as opposed to how many people would live, the preferred answer changed as well.

Civilization is a machine designed to organize and control human thought and behavior. Despite the increasing role of computer automation, human behavior is still the ultimate driver of the megamachine; human behavior is the ultimate source of many of our most pressing problems. It is also the ultimate source of our salvation. It is what we *do* that matters. It stands to reason that insight into solutions for the problem of civilization—if there are any solutions—can be gleaned from a close look at how human behavior is being directed, and at how it can be altered and redirected. Because human behavior is the critical issue, the science of psychology, the study of thought and behavior, offers a potentially useful way of framing the questions. Psychology can provide ways of thinking about how we are connected to the machine and processed into the system. It may also offer some potent clues to how we might disengage from and eventually dismantle the machine of civilization in the pursuit of more human—and humane—lifestyles.

The science of psychology provides a specific frame within which to organize our thinking about our situation in modern civilization. It is by no means the only frame. Nor is it better in any unqualified sense than other possible ways of framing our situation. Psychology as the science of behavior provides a way of organizing our thinking about both problems and potential solutions in terms of actual human behavior: what we do. Note that what we do is not always overt and observable. A lot of what we do is covert, in the form of reflective thought and non-conscious mental and emotional activity. It is important to keep in mind that a good deal of our behavior is not open to our conscious inspection. The methods of empirical psychology are designed specifically to provide insight into these facets of our behavior that are not necessarily obvious on the surface.

In the name of full disclosure, I need to mention that my professional training is in the area of cognitive psychology. Cognitive psychology is typically defined as the study of the acquisition, retention, retrieval, and use of knowledge. The psychological frame is one that tends toward mechanistic explanations of human behavior, explanations that treat human behavior as the output of mechanical processes. Cognitive psychology is a highly mechanistic sub discipline of psychology that makes extensive use of the computer metaphor. At some point along the way I became aware of the limitations of this approach, and I suspect that this awareness has something to do with the reasons I am writing this book. There are major domains of human experience in which mechanistic explanations are entirely mute. Consider the nature of human consciousness, for example. It is clear that consciousness is somehow related to brain activity, and according to the scientific party line the brain is an organ that, through neural activity that we don't yet quite understand, generates conscious experience. In other words, conscious experience is a function of some consciousness-producing mechanisms of the brain. I have heard this referred to as the *excretive theory* of consciousness, and it reflects the way that most psychologists (and most other people) think about the relationship between brain activity and conscious experience. However, there is an equally valid possibility. Instead of being something produced by brains, maybe consciousness is a fundamental feature of the cosmos and exists as a potentiality everywhere in the universe; rather than create consciousness, the brain acts like receiver tuned to a specific frequency of this universal consciousness. According to this view, we all have a unique conscious experience because our brains are slightly different from each other's and are therefore translating a different slice of universal consciousness. I doubt that any mainstream psychologists ascribe to anything like this second view despite the fact that, not only are both views equally mechanistic, but there is no principled way to distinguish between them. They make exactly the same empirical predictions. And note that the excretive theory is no more parsimonious than the receiver theory. It is just as impossible to imagine that conscious experience somehow mysteriously emerges out of nothingness from the electrochemical activity of neurons as it is to imagine it existing as a ubiquitous presence throughout the universe. I suspect the overwhelming preference for the excretive theory has something to do with a latent factory-production analogy that relates to the historical ties between science and industrial technology: the brain is a machine that assembles conscious experience for our consumption.

Humans are not computers, and to frame human thought and behavior solely in terms of mechanism is to miss something critical about what it means to be human. Nevertheless, the idea of mechanism makes for potent metaphor. In addition, we are talking about how human thought and behavior have become entrained to the workings of an actual machine, the megamachine of civilization, so a mechanistic psychological frame does not seem inappropriate. However, we need to be continually alert to the fact that when we talk about human beings in mechanical terms, we are employing a machine metaphor; we need to be alert to the limitations and framing effects involved. I want to encourage a practical approach to the use of a psychological frame as we proceed, an approach based on how the understanding or insight provided by psychology might be applied to specific concrete problems of civilization; in other words, how we might use psychological insight as a tool.

Using the Master's Tools

There is some not-so-subtle irony in using science—itself a product of civilization—as a tool to critique civilization, and the irony is not lost on me. It is actually a deep and abiding contradiction, a contradiction that lies at the heart of the main thesis of this book: as a human living under the oppressive directives of civilization, I am forced into a lifestyle that is counter to my very humanity. I am forced to live a life of contradiction, and like any other slave, if I am to resist by destroying the master's house, I have little choice but to use the master's tools to do so.

Philosophers of science trace a conceptual distinction between two perspectives regarding the relationship between scientific theories and reality. According to the perspective known as *realism,* scientific theories are actual descriptions (although perhaps imperfect) of underlying reality; that is, science tells us how the world really is. According to the perspective known as *instrumentalism,* scientific theories are useful for making predictions and for organizing our thinking about the world; science, according to an instrumentalist, doesn't tell us how the world really is, rather it is a useful tool that can be applied in a wide variety of ways. In what follows, I am assuming a strictly instrumentalist perspective. The science of psychology is a useful way of framing the issues and making predictions that can inform potential courses of action. I view realism as not only unwarranted given the extremely limited nature of human intelligence, but as potentially dangerous, a slippery slope into what Neil Postman, in his book *Technopoly,* calls *scientism.*

Postman (1992) defines scientism as consisting of three interrelated ideas: (1) that the methodology of the natural sciences can be applied to human behavior, (2) that society can be organized in a rational way through the application of social scientific principles, and (3) that science can serve as a comprehensive belief system that can satisfy all human needs including provide a meaning to life. I want to make it clear up front that I ascribe to only the first of these ideas, and even then not without strident caveats and considerable reservations. The methods of science provide a potentially useful frame. And it is possible to apply this frame to human behavior. But my intention in this book is openly subversive with respect to the next two core ideas of scientism. I am particularly antagonistic toward the second core idea; I find the idea of social engineering to be personally repulsive. Science is part of how the machine controls, and social science is the science of predicting and controlling human behavior. The evil of this is palpable. Who decides which behaviors to augment and in what ways and to what purposes? I want to expose how the machine of civilization employs the principles of social science within its operative programs and how, largely because of this, its operative programs are becoming increasingly oppressive. I have two overarching goals in this book that run counter to idea (2): I want to find ways that we might fortify ourselves against the machine's attempts at programming us; and I want to identify and target exploitable lynchpins in the system.

The third core idea of scientism, the notion that science can provide a comprehensive belief system, in addition to being pure hubris, completely misinterprets what kind of a thing science is. As the discussion of consciousness above highlights, mechanistic science can't even tell us how we should understand the source of our most basic personal experience: our personal awareness of experience itself. Science provides us with a very useful way of asking questions, but it is important to remember it is only a tool, and as with any tool the final result depends as much on the person using the tool and the purposes to which the tool is being applied as it does the nature of the tool itself. A hammer can be used to build a house or a gallows, and the best hammer in the world is no guarantee that a house built with it will stand or the gallows will hold fast when the rope snaps taut. But our purpose here is as much demolition as it is construction. And given the scope of the task at hand, I suspect we are going to need every tool at our disposal.

The Elephant in the Butterfly Net: A Fable

I will be incorporating a number of theories and empirical findings from the science of psychology into the chapters that follow. But I will be applying this research in a way that is different than it is normally applied. I am not trying to support or elaborate any one theoretical perspective. I am certainly not trying to advance the field of psychology in general or pioneer a new sub discipline. Instead, I will be looking at what the research has to say about civilization. Taken together, the data I discuss point to an undeniable and singular conclusion: that the continued existence of civilization is not in our best interests as individuals or as a species. I will also be exploring what the research suggests about how we might extricate ourselves from the omnipresent grip of the machine. It is a fairly straightforward matter to demonstrate ways that civilization is a problem for us. Deciding what to do about it is a different matter, however. To anticipate the conclusion that the data lead me toward, consider the following fable and its potential resolutions as an allegory.

Suppose there was an elephant who somehow managed to get a butterfly net stuck across the front part of the top of his head. Suppose further that the elephant knew that the purpose of a butterfly net was for capturing butterflies. Suspend your disbelief a bit further and suppose that the elephant thought that because it was caught in the net, it was, like a butterfly, trapped, and had no recourse but to submit to the demands of the person holding the other end of the net. And suppose that, tragically, there was in fact no one on the other end of the net. You are walking through the jungle and happen upon this elephant, which is by this time well on the way to starvation because, being trapped in the net, it has not been able to move from the spot for several days. What would you do to try to save the elephant? Feed the elephant by hand so that it doesn't die? Inform the elephant that there is no one holding the other end of the net? Attempt to convince the elephant that it is not a butterfly, and thus not subject to the rules of butterfly nets, that a butterfly net is powerless against its massive bulk? It seems the simplest solution might be just to remove the net.

There is one other possibility, however. Since the elephant is already convinced it is helpless and at the mercy of its captor, you might simply grab hold of the other end of the net yourself and start issuing commands.

PART 1: OUR NATURAL PLACE

Evolutionary Psychology

Behavior as Adaptation

Darwinian evolution is based on the principle of fitness. Fitness refers to an organism's capacity to navigate environmental challenges long enough to reproduce viable offspring. Fitness is a statistical characteristic; it's a matter of probabilities: the higher the likelihood of viable offspring, the higher the fitness. It is impossible to gage an organism's future fitness with any certainty, however, because it is impossible to predict with certainty what future environmental challenges will look like. A seemingly benign or even potentially detrimental genetic idiosyncrasy may provide a survival advantage when environmental demands change. Conversely, a clearly beneficial characteristic in one context, say, an oversized brain in a scavenging, savanna-dwelling primate, may become a one-way ticket to oblivion in a future context.

There are two key biological requirements for Darwinian evolution: heritable variation in the characteristics of individuals within a population and differential reproduction among those individuals. And there are two general mechanisms that drive evolutionary change: natural selection and sexual selection. Natural selection is the principle that some heritable characteristics provide a fitness advantage with respect to the challenges posed by the natural environment. Individuals whose body colorings allow them to better blend into the background, for example, are less likely to be victims of predation than individuals who blend less completely. Sexual selection explains the presence of behaviors and body structures that seem to be irrelevant from an environmental fitness standpoint, or appear actually to reduce an organism's fitness. The classic example is the male peacock's tail feathers, which reduce the ability to escape from predators. Fancy tail feathers, however, serve a signaling function that attracts female peacocks; and in terms of producing viable offspring, the increased ability to attract females offsets the reduction in the ability to avoid the leopard's jaws—at least for peacocks.

Species' populations tend toward equilibrium in terms of the presence and range of selection-relevant characteristics. Most extreme deviations are selected against. If environmental conditions remain unchanged for an extended period of time, the range of a species' characteristics will remain relatively stable as well. Populations can evolve fairly rapidly in response to rapid changes in environmental conditions, however. This pattern, long periods of little or no change sporadically interrupted by brief periods of rapid adaptive change, is called *punctuated equilibrium* (Gould & Eldredge, 1977). Note that a "brief period" on the evolutionary time-scale can be tens of thousands of years. Additionally, it is important to note that for any evolutionary change to occur, the triggering change in the environment has to involve changes in persistent or recurring features of the environment. A one-time (non-catastrophic) event or conditions that last only a few generations are usually not evolutionarily relevant. This requirement, in conjunction with the time frame involved, limits the potential evolutionary relevance of any feature of contemporary civilization, with the potential exception of the future relevance of wide-spread loss of habitat and the accumulation of long-term environmental toxins. It is also important to note in this context that when a species is unable to adapt to changes in environmental conditions (rapid or otherwise), the species ceases to exist.

Not every characteristic that is passed along from generation to generation is a result of evolution directly. "The filtering processes of natural and sexual selection result in three products: adaptions, by-products of adaptations, and random variation or noise" (Michalski & Shackelford, 2010, p. 511). Adaptions are structures or behavioral tendencies that are designed to accommodate features of ancestral environments, the pattern of color on a butterfly's wing that allows it to blend with the flowers on which it typically feeds, for example, or the human preference for calorie-dense foods to accommodate unpredictable periods of food scarcity. By-products are "tag-alongs" to adaptations, structures or behavioral tendencies that result from adaptations, but are themselves not directed at solving the environmental problem that generated the adaptation. These are sometimes referred to as *spandrels,* after the sweeping ornate structures in gothic architecture that serve no supportive function. Human eye color may be a spandrel (although some, e.g., Gardiner & Jackson [2010], have suggested that eye color may have played a role in sexual selection among ancestral Northern Europeans). Random variation or noise is a category that includes mutations or anomalies that are neutral in terms of their influence on fitness that get passed along because there are no countervailing selective pressures. Both by-products and originally neutral mutations can eventually become targets for selection if they help to solve a fitness-related problem for the organism.

Physical characteristics are not the only important consideration when it comes to fitness. Natural and sexual selection operate on biological structures, but a key driver in many situations is behavior; it is what an organism *does* that determines whether its genetic material gets passed along to the next generation. The role behavior plays in fitness increases with its complexity in a rich-gets-richer way; the more extensive the behavioral repertory, the greater the potential for adaptation. It goes without saying that the human behavioral repertory is extensive. And our capacity for adaptation is a matter of record: when European explorers set out to extend the dominion of their kings and queens, they found humans already occupying virtually every inhabitable terrestrial environment on the planet. Thus the question of what is it that makes us human is answered by referring not just to the evolved characteristics of our physical form (large brain, two legs, etc.), but to our adaptation-relevant behavioral propensities (social, complex tool manufacture, bipedal gate, linguistic communication, etc.)—a slight twist on the old saw about a duck: to be a duck it has to quack like a duck as well as look like one. The point is it's not just our physical biology that makes us human, it's our psychology as well; it's what we do and how we do it.

Biology is foundational, of course. Complex human behavior would not be possible without the evolution of a nervous system of comparable complexity. According to the *principle of proper mass* (Jerison, 1973), there is a direct relationship between the complexity of a given behavior and the amount of neural tissue necessary to support that behavior: the more complex the behavior, the more neural processing required; and the more neural processing required, the more brain tissue needed. The human brain is roughly two percent of body weight, which makes it, pound for pound, the largest brain of all primates, and the second largest of all vertebrates (we are beaten out by the feather-weight Etruscan shrew with a brain to body weight ratio of 3.3 percent). The structure and complexity of the human brain reflect a detailed evolutionary history of the species, as adaptations generated new modifications and extensions of old structures. Evolution doesn't generate new structure out of thin air; rather it works by modifying what is already there. Thus the complex behavioral capacities that we consider our defining human attributes result from the modification or extension of neural structures that likely supported a related but slightly different set of capacities in ancestor species. Our brain still caries remnants of its earliest configurations, from fish to amphibian, to mammal, to primate, and from Australopithecus to Homo erectus, to archaic Homo sapiens. It is all still there, and all of it playing a fundamental part in what makes us human.

Evolutionary Psychology

Evolutionary psychology is an area of theoretical speculation and empirical investigation that emerged in the latter part of the twentieth century, although, arguably, it dates back to Darwin himself. According to early spokespersons for this perspective (Cosmides, Tooby, & Barkow, 1992), the basic premises of evolutionary psychology include (1) there is a universal, pan-human human nature that exists at the psychological level, (2) that human psychology consists of mechanisms that evolved as adaptations to life in Pleistocene hunter gatherer band society, and (3) that cultural variability can provide insight into the nature of these mechanisms. The first premise, which is supported by the absence of consequential genetic differences among geographically disparate members of our species, nevertheless needs to be taken on faith, with the caveat that we will never be in a position to construct a comprehensive understanding of human nature because, being human, any understanding of human nature that we are able to construct will always be, in some sense, a product of that same human nature. The third premise turns traditional conceptions of the relation between psychology and culture on its ear. According to the evolutionary view, culture is an expression of evolved psychological mechanisms, and cultural variability reflects differences in the expression of these mechanisms that are due to geography, climate, population density, type and abundance of food sources: differences in the requirements for survival associated with different environmental circumstances.

"Theoretical and empirical work consistently shows that selection nearly always favors behavior and maximizes *individual* reproductive success, *regardless* of its implications for the welfare of the group of which the individual is a part" (Bird and O'Connell, 2006, p. 162, italics in original). Cultures evolve in the sense that they change over time—and can do so in a fairly dramatic and rapid fashion, but the resulting changes are always informed and constrained by underlying psychological mechanisms of individuals, mechanisms that change according to a far slower timeframe. And by looking at what varies and how from culture to culture (or from one historical period to another), and at what remains constant, we can get some feel for the nature of the psychological mechanisms involved. In an analogous fashion, a close comparison of superficially very distinct human languages has allowed us to map out the universal nature of language as an instinctive psychological capacity (Pinker, 1994).

The second premise of evolutionary psychology, the assumption that our contemporary psychological makeup is an adaptation to Paleolithic life-ways, is the one that is of primary importance to us here. The assumption that we are designed, biologically and psychologically, to be Pleistocene foragers demarcates both the source and the nature of the mismatch between our innate human predilections and the demands and expectations of life in the mechanical quasi-culture of global industrial civilization.

Evolutionary psychology is an integrated approach to psychology that explains human perceptual and mental programs and behavioral predispositions in terms of how these programs and predispositions have been selected for by evolutionary environments (Bereczkei, 2000). It is an integrated approach because it cuts across all of the sub disciplines of psychology, providing a framework that links developmental psychology, for example, with physiological, cognitive, and social psychology, showing how the phenomena of interest in these subareas are tied to the same sets of evolved programs. The word program is being used as something more than just analogy here. Evolutionary psychologists conceive of our psychical architecture as being composed of a large number of self-contained modules, something like the self-contained programs and "drivers" in a complex computer software system, each designed to cope with a particular adaptive problem. It would not be a complete mistake to think about these modules as instincts, although the way the term instinct is typically used doesn't capture the level of specificity involved with evolved modules. For example, what humans perceive as a unitary visual image is really the output of two distinct modular processing systems, one that provides us with color information and one that provides information about movement. These two visual processing systems operate in parallel and are independent of each other, despite the seamless nature of our visual experience. The self-contained nature of modules means that their operation is automatically triggered and they are largely unresponsive to top-down attempts at "reprogramming." We share many modules with other animals, but there are several that are unique to humans. Language, for example, is unique to humans, and is thought to be supported by a number of distinct modules; and it's the existence of these pan-human linguistic modules that accounts for the universal features of human language. Even apparent individual differences among members of the same culture, the variety and uniqueness reflected in individual personality differences, are due to the expression of shared modules. "The psychological mechanisms underlying personality have evolved over human evolutionary history because they solved the adaptive problems ancestral humans confronted" (Michalski & Shackelford, 2010, p. 512), and these domain-specific mechanisms are species-wide. There are universal features of human personality just like there are universal features of human language.

According to this view, the ways that we interpret both the physical and social world are dependent on the development of numerous adaptive modules. As an example, consider a ubiquitous feature of our social interactions: the *theory of mind,* the instinctive interpretation of agency that we apply to other people. We automatically interpret other people's actions in terms of internal goals and intentions despite any direct evidence—we obviously can't see thoughts and intentions. A rudimentary theory of mind develops in children by age 3, and becomes increasingly sophisticated as other cognitive functions mature (e.g., Hughes & Ensor, 2007). The application of a theory of mind is not limited to people; we frequently interpret animal behavior—and even machine behavior ("my computer is acting up again")—in terms of goals, desires, and so forth. It is easy to see the advantages a theory of mind offers in terms of predicting and understanding the behavior of others. What is not so easy to see, because its operation is automatic and largely invisible, is how it provides a fundamental scaffold supporting virtually all of our social experiences.

Culture plays a role in the development of a theory of mind in the same way that exposure to a specific spoken language is necessary for the development of our linguistic capacities. Before a theory of mind can be employed successfully, it is necessary to learn cultural norms and expectations with respect to goals, desires, and intentions. That is, there has to be a culturally-shared theory of why people do the specific kinds of things they do. Culturally shared explanatory frames of human behavior are called folk psychologies. Although different cultures have distinct folk psychologies, all are supported by an underlying theory of mind:

> People in different cultures may elaborate their folk psychologies in different ways, but the computational machinery that guides the development of their folk notions will be the same, and some of the notions developed will be the same as well. Humans come into the world with the tendency to organize their understanding of the actions of others in terms of beliefs and desires [i.e., with a theory of mind], just as they organize patterns in their two-dimensional retinal array under the assumption that the world is three-dimensional and that objects are permanent, bounded, and solid. (Tooby & Cosmides, 1992, p. 90)

It is interesting to speculate how the theory of mind informs—and, perhaps, misinforms—our understanding of our own personal behavior as well as the behavior of others. How often do we interpret actions in terms of rational agency when such an interpretation is unwarranted?

Emotion reactions can be seen as modular adaptations as well. Darwin pointed out that our emotional responses to environmental situations reflect our evolutionary past. Consider the case of phobias, irrational fears of specific objects or situations that are entirely out of proportion to the degree of threat actually present. People are far more likely to develop a phobic response to situations that are in fact potentially dangerous, and would have posed a potential danger to our distant ancestors. The most common phobias include the fear of heights, crowds, exposed or closed-in spaces, and potentially dangerous creatures such as spiders and snakes. The prevalence rates for *bicycle phobia* or *toaster phobia* are virtually zero despite the fact that interaction with either of these devices is far more likely statistically to lead to physical harm than interaction with the targets of many of the most common phobias. Contemporary researchers (Öhman & Mineka, 2001) argue for the modular nature of fear and fear learning. Phobic responses are an example of a highly adaptive fear response system. When it comes to potentially dangerous situations, it is better from a fitness standpoint to err on the side of too much fear rather than not enough. Additionally, acutely dangerous situations do not lend themselves to learning about though accumulated experience. The presence of an evolved fear module (or modules) solves the learning problem, but at the expense of being able to adjust the fear response to accommodate dramatic changes in environmental situations. Urban environments contain numerous acute dangers to which we have no evolutionary preparation. In addition, many of the characteristics of civilized life, for example the ubiquity of crowds, strangers, and both exposed and closed-in spaces, are situations very likely to trigger an automatic fear response. Again, the mismatch: much of the emotional stress associated with modern living can be traced to functional adaptive systems that are operating in an environment substantially different from ones for which they were designed. What other systems are likewise fully operational despite the fact that the situations they are designed to accommodate are no longer present in anything like the form in which they were present as the systems evolved? How have these systems, originally designed for life in hunter-gatherer band society, been re-appropriated or redirected by the demands of life in modern industrial civilization? Or, to turn the question around in anticipation of Part 3 of this book, how does the continued existence of modern civilization depend on the persistent misapplication of these evolved systems?

There is probably not a general human tendency that

cannot be couched in terms of its adaptive potential. Even tendencies that appear on the surface to be maladaptive may turn out on closer inspection to be a result of an ultimately adaptive process. For example, some have suggested that superstitious behavior, defined as behaving as if there is a cause and effect connection between two unrelated events, may have an adaptive function (Foster & Kokko, 2009). On first blush it doesn't make sense that seeing connections where none in fact exist (sometimes called *illusory correlation*) would offer any kind of survival advantage. Superstitious behavior is an example of what statisticians call *Type I error,* the mistake of deciding that a chance event is due to something other than chance. In an environment in which the consequences of making a *Type II error,* deciding that there is no connection between co-occurring events when there is in fact a cause and effect relationship, can be deadly, natural selection generates strategies that err on the side of making unwarranted connections. And so we are prone to forming superstitious beliefs, that a full moon has a mysterious power over events, or that day-by-day fluctuation in gasoline prices have something to do with national politics.

There has been considerable speculation about the possibility that religion has adaptive potential, and that the ubiquity of religion, defined in general terms as belief in the supernatural, or belief in counterintuitive agents, reflects a modular fitness-related adaptation (Sjöblom, 2007). It is not clear, however, whether religion is an adaptation, a byproduct, or something else entirely. The capacity for religion requires the presence of cognitive mechanisms that are themselves adaptations, the capacity for symbolic thought, for example, so it is easy to make the case that religion is a spandrel. Perhaps it is a byproduct of a too-broad application of the theory of mind, or simply another manifestation of the tendency to interpret co-occurrence in terms of cause and effect relationships that also produces superstitious behavior. Although there is some evidence of ritual during Paleolithic times, systematized ritualistic practices probably didn't play a dominant role in social life until the Neolithic. Neolithic domestication led, ironically, to an increased level of dependence on the caprice of natural forces. For band society, if the food was scarce in one area, they could easily pick up and move in search of better hunting grounds. The adoption of an agricultural lifestyle led to a dramatic reduction in mobility, and to a loss of flexibility in terms of dealing with challenges such as the loss of food due to drought, flood, or pestilence; and an increase in population density led to an increase in susceptibility to communicable disease and both intergroup and intra-group conflict. The greater dependence on the benevolence of unseen and uncontrolled powers that emerged during the Neolithic may have had a superstition-enhancing effect. An argument can be made that the presence and dominance of ritual and religion increases with dependency and is inversely related to autonomy, personal power, and control. Regardless of whether ritual and religion have adaptive functions, it is clear that the complex, compulsive, institutionalized religion that is a characteristic product of civilization is not an evolved adaptation, but a mechanism of behavioral control, a coercive tool of power. I have some more to say on this topic in Part 2.

Not surprisingly, human memory has also been found to reflect predictable adaptations to ancestral environmental conditions. The results of a set of studies conducted by Nairne, Pandeirada, Gregory, and Van Arsdall, (2009) provide evidence that our memory systems are organized in such a way that they more readily accommodate information relevant to survival in ancestral hunter-gatherer situations. Study participants were presented with a variety of survival scenarios, and then presented with a list of concrete nouns (e.g., chair, snow, orange) and asked to rate the nouns in terms of their relevance to the situation presented in the scenario. Three scenario conditions were used in their first experiment: hunter, gatherer, and scavenger hunt. Recall of the words was identical for both the gatherer and hunter conditions, and both of these conditions showed superior recall to the scavenger hunt condition. In a second experiment, words rated relative to a hunting-for- survival scenario were better recalled than the words rated relative to a hunting contest scenario. As an interesting aside, these researchers failed to find any memory-related gender differences. Males and females showed an equivalent memory benefit for hunter-gather and hunting-for survival conditions. Although it is admittedly a stretch to go from results of a single 21st century word-memory study to speculation about the general nature of Pleistocene social conditions, the failure to find gender differences is nonetheless consistent with anthropological and archaeological evidence that support the lack of a uniform strict gender-based division of labor in ancestral band society.

What other basic human capacities might be linked directly to our ancestral life-ways? Some have suggested that even such things as the capacity to hate may be an evolved response to inter-band resource competition (Waller, 2004). The tendency to categorize groups according to us versus them and the ubiquity of ethnocentrism and xenophobia are most certainly linked to our social evolution and our need to define the parameters of the tribe. A more subtle kind of adaptation to the demands of life in close-knit cooperative social groups is the capacity to regulate and control some of our less social predispositions, our predispositions for aggression and sexual gratification, for instance. That is, to live in cooperative society with other humans we need to have some context-sensitive way of overriding some of our other behavioral predispositions. We need to be able to engage in cost-benefit analyses with respect to our behavioral responses in social situations, and exert effortful control over many of our automatically triggered modular mechanisms (MacDonald, 2008). The ability to exercise effortful control, the ability to inhibit competing urges and desires and behaviors is an essential capacity for cooperative social engagement, one that distinguishes mature human adults from infantilized adults and children. In Part 2 we will explore how an important limitation in our evolved capacity to self-regulate has been exploited and directed in the service of the nonhuman goals of the megamachine.

Caveats and Complications

Ecological psychology (not to be confused with ecopsychology, which we will discuss below) is the name given to a perspective, also called *situation theory,* in which behavior is seen as a product of an ongoing complex interaction between affordances of the environment and abilities of the organism (Gibson, 1986; Greeno, 1994). *Affordances* are qualities of the environment that support or serve as prerequisites for the performance of specific activities or behaviors, and *abilities* are, similarly, qualities of the organism that support the physical production of specific activities and behaviors. A tree stump of a specific height and shape might afford sitting-upon, for example, and the ability to sit assumes that the organism is equipped with a particular kind of body structure. The central feature of the ecological psychology perspective is the interactive relationship between the organism and its environment. Context is everything. Abilities are never general; they are always constrained by specific features of the environment in which they are expressed. And the affordances of the environment are always relative to the goals and abilities of the organism. A tomato affords eating if I am hungry, and it affords throwing if I am angry about a poor stage performance. And my ability to consume or throw the tomato relies on capacities of my digestive system and learned coordination behaviors respectively. It is the ability-affordance interaction that is important.

Our psychological and perceptual systems evolved to operate upon affordances of the natural landscape (e.g., Orians & Heerwagen, 1992), most of which are not found in artificial cityscapes, and relative to goal-directed abilities that correspond to goals that are no longer being pursued. Technology-infused artificial urban spaces provide a far less complex environment in comparison to naturally occurring places.

> One way to begin thinking about the sacrifice of qualities is to notice that the success of technology is almost always the victory of artificial complexity over natural complexity. Technological structures and processes are built from complex systems in nature that have been altered is such a way that their

substance can be put to use. Alterations of this sort are
frequently made with the idea of making things
simpler than they had been before. [...] As technology
advances, a world of artificial structures replaces the
world of complex structures given in nature; thus
people no longer live in anything remotely resembling
a natural setting. (Winner, 1977, p. 210)

Try this simple demonstration. Sit in a room—any room in any
building—close your eyes, and document every different type
of sound that you hear for a 15 minute period of time. Then
find a natural location—any natural space out of doors
anywhere— and do the same thing. Then compare notes. The
concrete and asphalt center of a crowded city, with all of its
traffic noise and frenetic activity, is still a far simpler place
than anything the natural world has to offer. It follows that
abilities we develop relative to modern goals and the
affordances of the physical places of civilization are a meager
sampling of what we are capable of.

The idea of affordances highlights the complexity of the
organism-environment connection. Scientific theory is always
a simplification of reality. The theories, hypotheses, and
speculations generated by evolutionary psychology are no
exception. Even something as central to the evolutionary
approach as the role of DNA turns out to be far more complex
than theoretical science suggests. Not everyone recognizes that
the nature of science is to simplify. And although some
psychologists recognize that the mechanisms of genetic
inheritance that undergird biological evolution are considerably
more complicated than the standard neo-Darwinian account
suggests, they appear to be in the minority. DNA is not really a
blueprint or a kind of code, for one thing. To say so is just
useful metaphoric simplification. DNA is not the actual target
or "carrier" of the process of natural selection either. Rather it
is the complex and interactive developmental progressions that
DNA initiates and modulates that serve as the locus of
adaptation. "Developmental programs do not assemble an
organism of fixed design but rather a set of expressed
adaptations according to variables such as age, sex, and

circumstance-dependent design specifications" (Michalski & Shackelford, 2010, p. 512). Children aren't born with a theory of mind, for example. The theory of mind emerges as a distinct capacity in response to the maturation that occurs in numerous cognitive systems as they are fine-tuned by environmental feedback.

Lickliter and Honeycut (2003) offer what they call a *developmental dynamics* view in which natural selection targets not genes, not DNA itself, but the potential patterns of development that maturational programs and physiological systems in interaction with environmental input engender. Consider a subpopulation of a species of rodent that finds itself living in an environment in which the primary food source is hard-shelled seeds. Because bone-density is influenced by the frequency and degree of recurring stress during development, exposure to hard food when they are young causes the jawbones of these rodents to respond by increasing in size and density as they mature. Individuals who develop the thickest jaws have a fitness advantage over the others. What is being selected for in this subpopulation is not genes for large jaws, but rather a complex array of genetically-mediated processes supporting the capacity to respond physiologically to a specific set of environmental conditions (hard food) in functional ways as the animal matures. It is not specific genes for specific physical traits that are important, but the dynamic interaction between genetically-based maturational progressions (involving numerous disparate physical systems, each with their own set of genes) and environmental feedback.

Keep these big-jawed rodents in mind as we explore our own species' evolutionary environments and the mismatch between our own genetic expectations and civilization. We haven't inherited specific adaptations so much as we have inherited the ability to adapt in specific ways. The distinction is not a trivial one. It is the dynamic interplay between our genes and our environment—the ability to adapt to existing conditions—that is the crucial issue. The ability to navigate and adapt to the demands of civilization is something that is acquired during a person's lifetime, utilizing genetic programs put in place long before civilization happened. Our inherited characteristics lead us to respond in specific kinds of ways to the requirements of civilized life. And this fact has not been lost on the megamachine.

Human Roots: Paleopsychology

Exploding the Myth of the Primitive

There is a powerful and pervasive myth, frequently caricaturized in the media, that the people alive during the Pleistocene were "primitive" in the sense of being incomplete or not fully human: the fur clad caveman carrying a thick club and dragging his woman around by the hair. According to John Mohawk (1992), this myth is informative with respect to the conceit and arrogance characteristic of modern Western culture, but has absolutely no bearing on reality:

> According to these visions, hunter-gatherer society was generally prelingual. The people spoke a protolanguage of guttural utterances and childlike gestures. If they had any complex thoughts, they were able to suppress expression of them through clumsy behavior. They appeared, quite frankly, to have suffered brain damage. [...] [E]arly man is usually presented as a social undesirable, especially in relation to his ideas about male/female roles. [...] Why are we presented with a vision of the common ancestor which offers that, 50 or so generations ago, all our ancestors were essentially dolts? The answer, I suspect, lies in the way modern society wants to see itself. (pp. 206 – 209)

Perhaps the most famous articulation of this myth comes from Hobbes and his claim that life in a state of nature was "solitary, poor, nasty, brutish, and short." Human existence would be a "war of all against all" if it were not for the protective walls of civilization.

The evidence, both logical and empirical, against the media-sponsored Hobbesian view is overwhelming. First, life in a close-knit tribal community is anything but solitary. In fact, it might be argued that our present degree of personal alienation, the pseudo-autonomy, isolation, and loose and shifting social networks endemic in our corporate consumer society would make life solitary beyond endurance from the perspective of our Pleistocene ancestors. And recent trends with respect to social networking technology seem to be actually increasing social isolation (e.g., McPherson, Smith-Lovin, & Brashears., 2006). Second, modern-day hunter-gatherers lead exceedingly rich lives, if the ability to freely pursue personal goals and the capacity to provide for material needs are used as metrics. Stanley Diamond had this to say in his book *In Search of the Primitive* (1974):

> The ordinary member of primitive society participates in a much greater segment of his social economy than do individuals in archaic civilizations and technically sophisticated, modern civilizations. For example, the average Nama male is an expert hunter, a keen observer of nature, a craftsman who can make a kit bag of tools and weapons, a herder who knows the habits and needs of cattle, a direct participant in a variety of tribal rituals and ceremonies, and he is likely to be well-versed in the legends, tales and proverbs of his people (a similar list could be drawn up for the average Nama female). The average primitive, relative to his social environment and the level of science and technology achieved, is more accomplished, in the literal sense of that term, than are most civilized individuals. He participates more fully and directly in the cultural possibilities open to him, not as a consumer and not vicariously but as an actively engaged, complete person. (pp. 142-143)

And far from "nasty," life in these subsistence communities can be extremely pleasant, filled with celebration, singing and dancing, and an enormous amount of free time. If by brutish Hobbes meant a lack of sophistication or culture, then his comments simply reflect his Eurocentric bias. And, of course he was entirely ignorant of the findings of modern anthropology and archeology with respect to life-expectancy as well. The evidence also suggests that our Paleolithic forefathers and foremothers suffered less from chronic disease, ate far healthier diets, were taller and more physically robust than subsequent generations, were happier, were less aggressive, and were relatively more egalitarian—both in terms of gender equity and the lack of power-based class distinctions.

Perhaps part of the misconception is that stone-age lifestyles were somehow inadequate to the tasks faced by stone-age people. They used very simple tools, for one thing. They did not have industrial technology to provide them with the sophisticated devices that we have. We should remember that most of our sophisticated machines are useful only because of our deep dependence on industrial technology in the first place. The delusion of progress strikes again.

The media-enhanced Hobbesian myth of the primitive is a prevailing meme of civilization. The belief that primitive life was inferior is based on a logical mistake, a variation of a prevalent error in reasoning that philosophers call the *naturalistic fallacy*. The naturalistic fallacy refers to an idea, perhaps first articulated by Hume (1992/1739), that people tend to conflate what exists with what is good, that *is* is synonymous with *should be*. You are falling for the naturalistic fallacy anytime you think that the mere fact that a situation or state of affairs exists is justification for its continued existence, and, by extension, evidence of its superiority to other potential situations. From a psychological standpoint, the naturalistic fallacy is reflected in two related cognitive predilections: our strong tendency to prefer maintaining the status quo even if there are clearly superior alternatives, a predilection called *status quo bias* (e.g., Anderson, 2003), and a ubiquitous judgment heuristic known as the *existence bias,* where people assign goodness and value to a situation, event, or future outcome based on the belief that the situation, event, or outcome represents an existing state of affairs (Eidelman, Crandall, & Pattershall, 2009). The existence bias can lead to a devaluation of past conditions simply because they are no longer present. The naturalistic fallacy and its heuristic underpinnings are evident in the passive, unquestioning acceptance of the status quo, and in appeals to tradition for tradition's sake: "we should do it this way because that's the way we've always done it before." The mere fact that a person occupies a position of power is seen as providing legitimacy to that person's power, for example. Or, more to the present point, the mere fact that civilization exists is seen as clear evidence of civilization's superiority over the life-ways it has displaced.

The meme of cultural progress surfaced during a casual conversation I was having with a biologist acquaintance of mine not too long ago. The biologist made disparaging Hobbesian comments about primitive lifestyles of our Pleistocene predecessors, specifically referring to them as "non-advanced and thus lacking in intellectual sophistication." I responded by suggesting that it depends both on what you mean by intelligence and on how you define advancement. Tens of thousands of years in relative harmonious symbiosis with their environment seems quite advanced compared to the unsustainable approach taken by modern global industrial civilization. His retort—actually an emotional and visceral attack—honed in on my suggestion of symbiosis and harmony. He referred to the mass extinctions that accompanied the appearance of humans in the Americas as evidence that humans have always exploited their environment in ways that were ultimately destructive. We just destroy more efficiently now.

That last part is certainly true. But it is doubtful that the megafauna extinctions seen in the Americas around the time that humans crossed the Bering land bridge were caused directly by humans. Research strongly supports the idea that climate was the major contributing factor, and that if humans did play a causal role, it was simply as a result of their wedging themselves into a precariously balanced ecological niche rather than the result of innate pathological destructiveness (Pushkina & Raia, 2008; Ripple & Van Valkenburgh, 2010).[3] And besides, using large mammals as a primary food source is not a good survival strategy:

> Studies of hunting/gathering peoples show that to hunt
> big mammals exclusively is a bad strategy;
> generalized subsistence is more efficient and reliable;

and indiscriminant hunting is inefficient and goes against long term survival. Indeed, the proposed high predation of megafauna, such as elephants, among prehistoric peoples is extremely naïve if one considers the time and labor necessary for hunting large animals. [...] Just prior to human arrival in North America, about twelve species of megafauna vanished. Among them were a huge carnivorous bear, a gigantic lion, two genera of saber-toothed tigers, the jaguar, a cheetah, and the dire wolf. There is virtually no evidence, such as stone implements, of confrontation between humans and most of the extinct animals. [...] Little evidence exists, then, that humans were responsible for the extinctions that took place at the end of the Pleistocene. (Shepard, 1998, pp. 33-34)

Let's allow for the sake of argument that there was in fact a causal connection between the arrival of humans and the disappearance of several North American mammals. If so, then it is another example of what happens when an invasive species is introduced into an established ecosystem. This has no relevance to the intrinsically destructive nature of human behavior. What about the ecosystems of indigenous Africa? There have been hunter-gatherer peoples living in Africa since the beginning of people, right up to the present day—hundreds of thousands of years of unflinching environmental support for the foraging lifestyle. It was only with the advent of wide-spread domestication that the interaction with regional environments became truly unsustainable. What about Native Americans communities in the time since the large-mammal die-off? In areas where agrarianism was limited, harmonious sustainable lifestyles were maintained until the arrival of Europeans.

Maybe my biologist friend's response can be understood in terms of differences in our definition of "symbiosis" or "harmonious." When I think of harmoniousness or symbiosis with the natural environment, I do not mean to infer that life was all peaches and cream, only that there was a balanced interdependence and a tendency toward equilibrium whenever the balance was disrupted. When a local human population overtaxed local sources of food (as many surely did), they suffered starvation and death. Eventually balance was restored. The general lifestyle was sustainable and maintained in check by a natural give and take—that's what I mean by harmonious. Such a lifestyle can exist indefinitely. The shift from subsistence band society to domestic agriculture is not an advancement when you consider it brings with it (1) a linear exploitation of the environment that is ultimately unsustainable, (2) division of labor accompanied by the hierarchical distribution of power (read: access to the resources necessary for survival), and (3) a three- or four-fold increase in the amount of work for the majority of the population (defined as time devoted to the provision of necessities—hunter-gatherers spend as little as three to four hours a day "working," and rarely if ever make a clear distinction between work and other kinds of activities). And industrialization represents an exponential compounding of the most negative features of the shift to agrarianism.

And, of course, the idea of "advancement" is tied directly to the delusion of progress, itself a by-product of industrial civilization. But beneath my biologist friend's response lurks a hidden implication that serves as a far more damning critique of humanity: the implication that the damage caused by modern industrial civilization is an emergent property of human nature. Humans are destructive creatures by design. We can't help ourselves; it's built into our DNA, part of who we are as a species. Termites build termite mounds, humans build planet-devouring civilizations. If this is true, then we are helpless to change our situation. We, along with the rest of the planet, are doomed to suffer the consequences of our evolutionary defects. We will revisit this issue below.

Foraging Culture

Civilization is not the same as culture. Civilization is not a particular kind of human culture, or some broad amalgam of cultural practices. Calling modern industrial civilization a kind of culture because, like culture, it provides a meaning-rich organizational framework for our thoughts and activities, is like calling a space shuttle a kind of packhorse because it allows us to transport things over a distance. And I balk at calling civilization a meta-culture for reasons that should be already apparent. Civilization does not serve as some kind of organizing principle for the coordination of disparate cultural traditions; it destroys them, eliminates them, and reduces any residual traces to superficialities of dress, diet, or speech. Civilization is at best a quasi-culture. Civilization is a machine that eventually makes all culture into a hollow shell, a superfluous appendage. Culture, true culture, is anathema to civilization.

All cultures evolve in response to the demands and opportunities of specific environments. That is, in essence, the purpose and function of culture: to facilitate environmental adaptation. But civilization is not just a product of cultural evolution. Civilization is the implementation of a pervasive and infiltrating mechanical power hierarchy that seeks to replace the need for culture. Civilization doesn't adapt to the environment. Civilization is a way of altering the environment, forcing the environment to change according to the whims of those at the top of the power hierarchy—for the singular purpose of supporting the continued expansion of the power hierarchy itself.

The mechanisms of cultural evolution can be seen in the changing patterns of foraging behavior in response to changes in food availability and changes in population density. Archaeological analyses suggest that there is a predictable pattern of dietary choice that emerges from the interaction among population density, relative abundance of preferred food sources, and factors that relate to the search and handling of various foods (Bird & O'Connell, 2006). In general, diets become more varied, or broaden, as population increases and the preferred food becomes more difficult to obtain. When a preferred food source is abundant, the calories in the diet may consist largely of that one particular food. But as the food source becomes more difficult to obtain, less preferable foods will be included and the diet will broaden. Such dietary changes imply changes in patterns of behavior within the community—changes of culture.

Behavior ecologists and anthropologists have partitioned the foraging process into two components with respect to the cost-benefit analysis associated with dietary decisions: search and handling (Bird & O'Connell, 2006). The search component of the cost-benefit ledger refers to the amount of work per calorie payoff (and other benefits such as the potential for enhanced social standing) associated with a food item's abundance, distance, terrain, proximity of another group's territory, water sources, etc. The handling component refers to the work per calorie payoff associated with getting the food into a state (location, form, etc.) in which it can be consumed. Search and handling considerations can be largely independent of each other. The residential permanence involved with the incorporation of agriculture reduces the search consideration greatly, and makes handling the primary consideration. Global industrial food economies change entirely the nature of both search and handling: handling in industrial society—from the perspective of the individual and the individual's decision processes—is reduced largely to considerations of speed and convenience. The search component has been re-appropriated and refocused by corporate marketing, and reduced to something called *shopping.*

Domestication, hands down the most dramatic and far-reaching example of cultural evolution, emerges originally as a response to scarcity that is tied to a lack of mobility and an increase in population density. Domestication is a way of further broadening the diet when other local sources of food are already being maximally exploited. Initial experimentation with animal domestication "occurred in situations where forager diets were already quite broad and where the principle goal of domestication was the production of milk, an exercise that made otherwise unusable plants or plant parts available for human consumption…" (Bird & O'Connell, 2006, p. 152). The transition to life-ways based even partially on domestication has some counter-intuitive technological ramifications as well.

This leads to a further point about *efficiency*. It is often said that the adoption of more expensive subsistence technology marks an improvement in this aspect of food procurement: better tools make the process more efficient. This is true in the sense that such technology often enables its users to extract more nutrients per unit weight of resource processed or area of land harvested. If, on the other hand, the key criterion is the cost/benefit ratio, the rate of nutrient gained relative to the effort needed to acquire it, then the use of more expensive tools will often be associated with declines in subsistence efficiency. Increased investment in handling associated with the use of high-cost projectile weapons, in plant foods that require extensive tech-related processing, and in more intensive agriculture all illustrate this point. (Bird & O'Connell, 2006, p. 153)

In modern times, thanks to the advent of—and supportive propaganda associated with—factory industrial agriculture, farming is coupled with ideas of plentitude and caloric abundance. However, in the absence of fossil energy and petroleum-based chemical fortification, farming is expensive in terms of the calories produced as a function of the amount of work involved. For example, "farmers grinding corn with hand-held stone tools can earn no more than about 1800 kcal per hour of total effort devoted to farming, and this from the *least* expensive cultivation technique" (Bird & O'Connell, 2006, p. 151, italics in original.). A successful fishing or bison hunting expedition is orders of magnitude more efficient in terms of the ratio of calories expended to calories obtained.

Of course, the members of foraging cultures have little or no explicit concern with efficiency per se. The modern idea of efficiency is, after all, a product of the industrial revolution. In fact, much of what occurs in day to day life in a subsistence culture might appear enormously wasteful from an efficiency-obsessed modern mechanical production perspective. But natural systems do not tend toward unnecessary complexity, so any appearance of wastefulness in human forager behavior reflects a failure to recognize the behavior's true purposes. Some apparently wasteful behavior can be explained by its "signal value" (Bird & O'Connell, 2006). Stotting in gazelles is a paradigmatic example of signaling behavior that appears on the surface to be wasteful, but in fact serves fitness-related purposes. Instead of running immediately on approach of a carnivore such as a lion or wolf, a gazelle will take a few steps and then leap high in the air as a way of signaling its strength and health to the predator, a way of sending the message "I am healthy and strong and I will be hard for you to catch, so don't waste your energy on me." Stotting also serves a potential signaling function to other gazelles, in the manner of the peacock's tail feathers, a kind of self-handicapping that provides evidence of good genes for potential mates. For humans in hunter-gatherer societies, engaging in big game hunting may serve more of a signal function than as a way of providing food. Males might engage in big game hunting because of its signal value even when the calorie payoff (perhaps less than one carcass per 30 days of hunting effort) is extremely small.

> To the degree the hunter is successful, two ends are achieved. First, because big-game hunting is a risky, skill-intensive undertaking, the good hunter marks himself as a powerful ally and dangerous adversary. His relationships with others are likely to be structured accordingly. Equally important, his successes make available a "public good," one that is of interest to all, unpredictably acquired, readily divisible, and thus likely to be shared widely [...], considerations that draw still more favorable attention his way. That attention might include deference to his wishes, support in disputes, positive dealings with his spouse and children, and more frequent mating opportunities. (Bird & O'Connell, 2006, p. 165)

The potent signal value of big game hunting might explain why Neanderthals were out-competed by humans: Neanderthals engaged in risky close-in spear hunting of large game and were reluctant to adopt other, less-risky means (e.g., atlatl, bow and arrow) because to do so would reduce mate choice opportunities (Bird & O'Connell, 2006). Some (e.g. Diamond, 1991) have even suggested that much of the reckless and unhealthy activity in which modern human males are prone to engage can be understood as self-handicapping behavior that serves a signaling function much like that of the peacock's tail or the bowerbird's labor-intensive bower-building. Young males are no longer able to prove their worth by bagging an auroch, so they join street gangs, smoke, drink excessively, used dangerous drugs, drive recklessly, or play contact sports.

Material displays, various artistic flourishes and ornamentation on weapons, jewelry—perhaps the development of art itself—may also have a signal-value explanation. The dramatic increase in the presence of symbolic artifacts that occurred 50,000 years ago may have nothing to do with any change in human cognitive ability such as an increased capacity for symbolic thought, but instead reflect an increase in human population density and the concomitant competition for resources that led to an increase in material display as a fitness signal (Bird & O'Connell, 2006). In other words, humans may have had the linguistic, symbolic, and cognitive representational capacity to produce a Chauvet or a Lascaux for hundreds of millennia before the social conditions were ripe for their expression.

This latter speculation about the emergence of symbolic art underscores an important point about the difficulty with inferring behavioral and intellectual capacities from actual behavior, sometimes referred to as the *competence-performance* distinction. There is almost always a vast gulf between what a person actually does (performance) and what he or she has the potential to do (competence) given the right contextual supports. For example, it has been estimated that the typical college graduate has a receptive vocabulary of between 40,000 and 50,000 words, but actually uses only about 4,000 or 5,000 different words when writing and only about 1,000 in daily conversation. It has commonly been assumed that the increase in artifact diversity and the relatively sudden appearance of art during the upper Paleolithic reflect an underlying change in human psychological capacities, an increase in linguistic acumen and symbolic ability, for example. Unfortunately, neither symbolic thought nor spoken language leave fossil traces. And a more parsimonious explanation is that the capacity for both symbolic thought and artistic representation have been latent potentialities since Homo sapiens emerged as a distinct species, but that these capacities did not manifest in ways that are detectable archaeologically until the right conditions for their expression appeared. Imagine an archaeologist 50,000 years in the future looking at the explosion in digital mass technology that occurred in the early 21st century as evidence of the sudden emergence of a new cognitive ability to multitask.

Because we did not possess a given tool or demonstrate a particular conceptual capacity sometime in the past is not proof of a limitation. Nor is the presence of sophisticated tools and capacities evidence of superiority relative to the past. The complexities of modern civilization require that we carry a different kit bag of tools, both actual and psychological, than those needed for life in a subsistence foraging society, but our tool-building capacities evolved to meet the daily needs of a foraging lifestyle, needs that are in many ways qualitatively different from the daily requirements of life within the grinding gears of civilization. We turn now to a discussion of some important psychological ramifications of this.

Evolutionary Psychopathology

Biophilia

Biophilia is a term Harvard biologist E. O. Wilson (1984) came up with to describe our evolution-derived affinity for living things and life processes. According to Wilson, natural selection has instilled in us a natural sensitivity for the natural world. Our ancestors depended on a keen awareness of nature for survival. And so even today, especially with many of us living in highly artificial environments, we still feel the pull of the natural. We enjoy scenic vistas and visit national parks and tropical vacations spots. We keep houseplants and pets. We adorn our walls with pictures of nature scenes. We surround our houses and shopping centers with trees, bushes, and lawns. All of these can be seen as manifestations of biophilia. Empirical support for Wilson's biophilia hypothesis abounds. Gullone (2000) reviews several categories of evidence, including a preference for certain landscape characteristics, the general aesthetic preference for natural versus urban scenes, and the stress reduction, mood enhancement, and positive cognitive changes that follow exposure to natural settings, wilderness views, and animals.

Biophilia is consistent with many of the ideas of evolutionary psychology. Psychological modules developed in response to the demands of specific environmental circumstances, and their functioning is triggered and supported by the presence of these circumstances. This suggests that there is more to biophilia than just a tendency for us to enjoy nature. Exposure to nature may be a requisite for the vigorous healthy functioning of our physical and psychological systems. Empirical research supports the contention that our affinity for the natural world is not just an interesting remnant of our evolutionary trajectory; nature-exposure appears to be essential for our physical and mental health. Researchers have found that even small periods of time spent in natural environments can have positive benefits with respect to numerous indicators of physical and mental wellbeing. Well over a hundred studies have found a positive relationship between spending time in nature and stress reduction (e.g., Frumkin, 2001). Horticulture therapy and pet therapy have been popular for years, and have proved especially useful for treating anxiety and depression (e.g., Kim, 2003; Moretti, et al., 2011). Studies in hospitals have found faster surgery recovery times for patients in rooms with a view of trees versus those in rooms with a view of a brick wall (Ulrich, 1984). People who are exposed to views of natural landscapes following stressful experiences calm down more rapidly (e.g., Kahn, et al., 2008). Nature exposure also appears to relieve the symptoms of ADHD in children—in some instances completely eliminating the need for medication (Kuo & Taylor, 2004).

As with evolutionary psychology in general, the idea of biophilia suggests a critical interplay between human nature and the rest of nature. The problem is that the natural world is changing rapidly, it is becoming increasingly unlike the kind of world in which our human nature evolved. It has been suggested that that recent planetary changes caused by human activity have been of such a magnitude as to warrant the addition of new epoch to the geological calendar: the Anthropocene (Steffen, Crutzen, & McNeill, 2007). Anthropogenic changes in atmosphere, climate, seawater, and land erosion patterns, as well as sweeping species extinctions mean that for all intents and purposes the Holocene ended with the agricultural revolution. I would argue that the changes caused by agriculture and more recently fueled by industrial civilization warrant not just the coining of a new epoch, but the unveiling of an entirely new—albeit potentially short-lived— era. We have entered the Anthropozoic. The differences between before and after the agricultural revolution are easily on par with the differences that straddle the two sides of the K-T boundary—and global industrial civilization is turning out to be the mother of all meteor strikes!

It is not just the physical environment that has changed. Civilization has also generated a dramatic change in the entire range of behaviors, expectations, goals, aspirations, and desires that give life its meaning.

> During the long course of human evolution, we valued
> nature and living diversity because of the adaptive
> benefits it offered us physically, emotionally, and
> intellectually. And people continue to need rich and
> textured relationships with natural diversity in order to
> achieve lives replete with meaning and value. (Kellert,
> 1997, p. 3)

The mismatch between our evolved predilections and industrial civilization is a knife with two edges. The first prods us to adopt mechanical life-ways that are alien from the perspective of our organic Pleistocene genes. The second cuts us off from the pursuit of meaningful goals and offers us surrogate activities whose shape and form promise fulfillment of our inborn needs but never truly deliver on their promise.

As the natural environment continues to change at an accelerated pace, as more species become extinct and more natural spaces succumb to the incessant grinding of the machine of global civilization, our ability to satisfy our biophilic drives will likewise continue to diminish. There is already good evidence that the bar is being lowered with each passing generation, a kind of *generational amnesia* (Kahn, Severson, & Ruckert, 2009) with respect to what counts as a rich and fulfilling human life. Technological "advance" is the prime mover of this tendency. Actual experience with the natural world is being rapidly replaced by experience with technological simulacra, or "technological nature." Our experience with nature is more and more mediated and ersatz. And the baseline for what counts as a meaningful nature experience is shifting with each generation. Kahn, et al (2009) conducted a set of studies in which the calming effects (heart-rate reduction following a stressful situation) found with prior research were compared between a room with a window looking out onto a natural setting and a room equipped with a high-definition plasma screen television showing the same scene. They found the typical calming results of nature exposure with the window, but the plasma screen was no different than a room with a blank wall. Even so, participants preferred the plasma picture to nothing at all. The authors believe that we as a species will adapt to the loss of actual nature and its replacement with technological simulacra, but there will by a substantial price to pay in terms of our physiological and psychological well-being.

> This tide of blight and devastation must be reversed if only for selfish reasons. The notion of biophilia emphasizes that healthy and diverse natural systems represent less a luxury than the potential for helping us realize lives of satisfaction and meaning. Celebrating our connections with nature inevitably renders our existence richer and more rewarding. Natural splendor is still the crucible in which our physical and mental well-being is forged. (Kellert, 1997, p. 4)

The idea of collective amnesia is a potent one. It suggests something lost, but also suggests the possibility of recovering what was lost, or at the very least, of relearning what we once knew. However, there are some things that, once lost, can never be retrieved. Species extinction is one way in which many of our potentially rich connections with the natural world are being permanently, irretrievably, forgotten from one generation to the next.

> The experience of extinction reveals, gravely, a related phenomenon, namely, the extinction of experience: that is, when species die forever our human experience is diminished—some types of experience actually become extinct, never to exist again—and equally, the experience and healthy existence of the larger ecosystem are diminished [...] Today there is rightly great concern about the decimation of the gene pool due to loss of biodiversity. Similarly, losing the diversity of experiences and interrelationships with nature is a decimation of the psychological and cultural "gene pool:" We become less fully human and the

ecosystem becomes less fully itself. (Adams, 2005, pp. 277-278)

In addition, it is possible that there are capacities and abilities that were prevalent in our Pleistocene relatives but are no longer present today: olfactory acuity, for example, and perhaps even more subtle powers of perception. It has been suggested that our Paleolithic ancestors had powers of empathetic identification with the other inhabitants of their world. "Interspecies empathetic identification may have been the norm in the paleolithic world, a capacity that has become atrophied through disuse in humans" (Metzner, 1999, p. 84). If so, then modern humans are suffering from a kind of pathological condition analogous to autism with respect to their relationship to the natural world (Berry, 1988; Metzner, 1999), and we have become quite literally blind to significant avenues of interactive potential.

Civilization as Psychopathology

The notion of pathology provides a useful frame for highlighting the effects of civilization's operation. Industrial civilization is a potent systemic physical toxin adversely affecting every organ tissue in the human body while it poisons every ecological niche in the biosphere. And all civilization—industrial and otherwise—serves as a chronic source of psychological stress for everyone yoked to the megamachine's flywheel. The idea of pathology presumes healthy functioning as a baseline, and that maladaptive functioning is evidence of deviation from this baseline. Applying an evolutionary framework to psychological disorders, an *evolutionary psychopathology,* suggests a dramatically different interpretation of what this baseline deviation entails:

In psychology and psychiatry, the terms pathology and psychopathology are most commonly used as references to diseases or disorders, the common conceptualization being that psychopathology is related to non-adaptive errors, malfunctions or breakdowns. In contrast, within the evolutionary psychopathology framework, it is proposed that many psychological states, currently identified as pathologies, may in fact represent the activation or manifestation of once adaptive strategies (i.e. application of biologically prepared rules in a relevant environment). (Gullone, 2000, p. 310)

That is, it is not a breakdown of normal psychological functioning that is causing the problem; rather normal functioning *itself* is the problem. Our psychological systems are operating in exactly the way they are designed, but these ways of operating are no longer functional within the contexts provided by civilization. Psychopathology is a predictable result of this mismatch.

Has our modern lifestyle, brought about by industrialisation, advanced technology, and corresponding cultural changes, enhanced our psychological well-being? [The empirical evidence] strongly indicates that it has not. Rather, the opposite conclusion may be warranted. Our modern lifestyle manifests as a large discrepancy between who we are and how we live. There are indications that this discrepancy may well be responsible for the increases in psychopathology evidenced in the modern world. (Gullone, 2000, p. 311)

There are two lines of empirical support for the evolutionary psychopathology interpretation: (1) evolution-inspired psychological research finding, among other results, evidence for evolutionary *preparedness* with respect to phobias, improved mood and cognitive functioning following exposure to natural settings, and enhanced social interaction in the presence of animals; and (2) studies finding substantial rural versus urban and cross-cultural differences in two of the most disabling psychopathologies, depression and schizophrenia (Gullone, 2000; Peen & Dekkar, 2004;

Sundquist, Frank, & Sundquist, 2004).

Cross-cultural variability in depression and schizophrenia is powerful evidence that lifestyle plays an important role in the expression of psychopathy, and suggest that there are degrees of mismatch with respect to our evolutionary expectations and the demands of society. But when it comes to mental health, the mismatch problem may be more complex yet. Humans don't come into the world fully formed, but are the result of an extensive and complex developmental process. Paul Shepard (1982) claims that the core of the mismatch problem is arrested development, a debilitating immaturity caused by obstructions to our natural epigenetic trajectory. Healthy psychological maturation depends on a life embedded in the natural world, a life in which the land and its nonhuman denizens play an essential role in the developmental formation of our essential psychological capacities. The insecurity, impulsiveness, anxiety, alienation, and childishness that define the 21st century civilized adult are a result of unmet inborn developmental expectations during childhood and adolescence.

> An ecologically harmonious sense of self and world is not the outcome of rational choices. It is the inherent possession of everyone; it is latent in the organism, in the interaction of the genome and early experience. The phases of such early experiences, or epigenesis, are the legacy of an evolutionary past in which human and nonhuman achieved a healthy rapport. Recent societies have contorted that sequence, have elicited and perpetuated immature and inappropriate responses. The societies are themselves the product of such amputations, and so are their uses and abuses of the earth. (p. 128)

Development occurs in stages defined by a series of overlapping *critical periods*, windows of time in which the maturing organism requires a specific kind of input or interactive feedback from the environment. Once the critical period for a specific set of capacities or characteristics has passed, the window closes; and if the appropriate environmental input was unavailable, the organism is left underdeveloped and less prepared to navigate future developmental challenges, and less likely to reach a healthy level of maturity as an adult. Civilized society not only reduces and distorts contact with the natural world—the environment in which our developmental programs evolved—but it also imposes demands and expectations that largely ignore the natural developmental progression, treating maturation as if it were a simple timeline, an accretive linear sequence rather than a cyclical and expanding process of person-environment interaction. Nostalgia for childhood, according to Shepard, does not represent a longing for our lost innocence or a desire for a happier or more carefree existence. It is a desire to return to a point where a healthy future adult self was still a latent potential.

The recently emerging field of ecopsychology, which has been described as an integration of psychology and ecology (Davis & Atkins, 2004), makes an explicit connection between modern mental health problems and the lack of contact with the natural world. Ecopsychology has a close affinity with environmentalism and environmental concerns, and sees our most pressing environmental problems and the decline in mental health associated with modern consumer culture to be manifestations of the same underlying pathology. Historian Theodore Roszak, an important spokesperson for the field of ecopsychology, coined the term "ecological unconscious" (after the Jungian notion of collective unconscious), defined as "our inherited sense of loyalty to the planet" (Rozak, 1992). And we ignore our inborn obligation to nature at our own mental peril. Roszak claims that "believing that we have no ethical obligation to our planetary home [is] the epidemic psychosis of our time." One of the primary aims of ecotherapy is to counteract what has been referred to as a "fundamental delusion of humankind" (Adams, 2005), the notion that we are somehow separate from nature. The dualistic human-versus-nature perspective promoted in Western society can be traced at least as far back as Descartes' mechanistic view of natural processes as opposed to the spiritual essence of the human soul/mind. The antagonistic flavor of the relationship between humans and nature is reflected in ancient Judeo-Christian cosmology (Shepard, 1982), and probably has its ultimate source in the emergence of large-scale domestication with the agricultural revolution. It is beyond the scope of this book to outline the conceptual ins and outs of the historical path of dualistic thought with respect to the natural would. It is enough to say that it is clearly not a tenable perspective.

According to Will Adams, a psychologist at Duquesne University, there are two consequences of realizing the non-dual nature of our relationship with nature: (1) we are less likely to exploit or abuse nature (it's part of us), and (2) we have a more comprehensive understanding of who we really are in the world. Because of our estrangement from nature, we are continually trying to find ways to reestablish contact. We are aching for closer contact, and this sometimes comes out in unhealthy ways. As an example of this unhealthy contact, Adams relates a story of man who shot a bald eagle simply because "you don't see too many of those birds around here these days." What is healthy for people also turns out to be healthy for the planet.

> Psychologists often emphasize that our relations with others may bring forth health or pathology, for both our self and others. Likewise research is revealing that this is also true in our relations with the natural world. Human well-being and the well-being of the natural world are mutually dependent. (Adams, 2005, p. 269)

Unfortunately, attempts to resolve the human-nature relationship by absorbing humanity within the natural world frequently ignore a real and crucial distinction between civilization as a maladaptive artificial mechanical process and human activity as natural and adaptive. This failure to distinguish the natural results of collective human behavior from the *un-nature-al* results of the implementation of a specific kind of centralized behavior-control system (i.e., civilization) is particularly apparent in discussions of our global ecological crisis. Various organismic metaphors have been used to highlight the role of human activity in the global ecological crisis in a way that makes civilization seem like a natural byproduct of human activity or an emergent property of human nature. For example, the exploding human population is sometimes compared to a cancer or a planet-wide parasitic infestation (Metzner, 1999). These metaphors also serve to obscure the true source of the problem: the machine of civilization. Remember, "machine" is not a metaphor in this context—civilization is an actual machine, not a metaphoric one.

It has become increasingly difficult to ignore the extent to which human activity is threatening the integrity of the biosphere. Global climate change, deforestation, overpopulation, a weakening ozone layer, widespread accumulation of environmental toxins, and an accelerating rate of species extinction are all results of human activity. However, the true locus of these problems is almost always misplaced. The source of our environmental woes is never pinned directly on civilization itself, but on some feature or features of human nature: our environmental problems are side effects of our alienation, a crisis of "consciousness and culture" (Adams, 2005); our problems are not due to the accelerating influence of a system of power and control that is becoming more and more efficient at precisely what it was designed to do. Recall my biologist friend's claim that humans are naturally destructive as a species. It's not civilization itself, but some either transient or inherent set of human flaws. For example:

> It is widely agreed that the global ecological crisis which confronts the world today is one of the most critical turning points that human civilization has ever faced. Furthermore, the realization is spreading that the root causes of environmental destruction lie in human psychology—in certain distorted perceptions, attitudes, and values that modern humans have come to hold. (Metzner, 1999, p. 80)

Global climate change is caused by distorted perceptions,

attitudes, and values? So all we have to do is lose the bad attitude and get with the program? Reminds me of a lecture my high school vice principal once gave me. The first sentence in the quote above highlights the true distortion: our global crisis is something that *civilization* faces. Civilization is the problem, but to hang the onus on civilization itself, squarely where it belongs, means that we would have to openly acknowledge what is really going on. We would have to see civilization for what it is and, most uncomfortably, recognize the nature of our dependency and the true extent to which we have become mindless slaves to the machine. The global environmental crisis is not the problem. It never has been. Civilization is the problem. There is no way to solve our environmental problems and at the same time allow civilization to continue.

The fundamental delusion of separation is necessary for the machine to function. A primary way that modern civilization exacerbates the delusion of separation is through technological mediation. More and more of our interactions with the natural world are being given over to technology. What results is an increasing separation, a growing distance between our actions and their real-world consequences. Technology, rather than making things better for us, is actually deepening our estrangement and impoverishing our experience in a variety of critical ways. It is to these issues we now turn.

PART 2: THE INVISIBLE MACHINE

Homo-Technologicus

Tools versus Technology

I want to start this section by exploring, in a somewhat belabored way, the hazy distinction between *tools* and *technology*. Quite often the two terms are used interchangeably. Perhaps more frequently *technology* is used to refer to tools in the aggregate, or to a set of instruments and techniques. But technology seems to me to be something beyond just a collection of tools and techniques. The difference between tools and technology seems to be a difference in kind. I don't want to get bogged down in a fine-grained, historically-supported, linguistically-informed semantic analysis, however; my goal here is merely to endorse a potentially useful distinction between tools and technology, one that is grounded in psychological reality.

We are a tool-using species, but humans are not the only creatures who make and use tools. Several other primates, for example, chimps, orangutans, capuchins, and macaques, have been observed using a variety of tools for collecting insects, cracking nuts, and even for fishing and hunting other mammals. A few bird species also make and use tools. New Caledonian crows are famous for their manufacture and use of barbed and hooked twig-tools. Although animal tool use is sometimes dismissed as "mere instinct," both bird and primate tool use appears to be acquired and maintained through social learning (Hotzhaider, Hunt, & Gray, 2010). Human tool use, however, is orders of magnitude more frequent, more sophisticated, and more broadly applied than that seen with any other species. Humans appear to be the only creature to "spontaneously and almost systematically use tools to modify their way of interacting with the world" (Osiurak, Jarry, & Le Gall, 2010, p. 517). Humans use tools for food, transportation, communication, comfort, entertainment, protection, distraction, and an open-ended list of other purposes; and humans are apparently the only creature to use tools to make other tools (Osiurak, Jarry, & Le Gall, 2010). Although other animals make and use tools, they do not have technology. Neither did Paleolithic humans, according to the characterization of technology that I endorse below: technology emerges as a product of the specialization (isolation) of knowledge (expertise) and the division of labor that are requisites for a domestication-based lifestyle.

So what exactly is a tool? Merriam-Webster's online dictionary offers three definitions of tool that are relevant to us here. First, a tool is "a handheld device that aids in the accomplishing of a task," "something (an instrument or apparatus) used in performing an operation or necessary in the practice of a vocation or profession," and "a means to an end." Scientific and academic definitions of tool, although typically more precise, are largely based on convenience rather than psychological reality (Osiurak, Jarry, & Le Gall, 2010); nevertheless, there is a high degree of consensus among academic definitions with respect to three essential features: (1) tools are discrete objects in the environment that (2) amplify the sensorimotor capabilities of the user and (3) are limited to what is actually manipulated by the user (Osiurak, Jarry, & Le Gall, 2010, p. 518). A pen, by this account, is clearly a tool. It is a discrete object that amplifies the user's ability to make marks on surfaces, it requires no parts external to the object, and the process involves no external input or activity that is not under the user's manipulative control. Merriam Webster's also offers three ways of defining *technology:* "the practical application of knowledge," "a manner of accomplishing a task using technical processes, methods, or knowledge," and "the specialized aspects of a particular field." *Tool* and *technology,* are frequently used in overlapping and quasi-synonymous ways; and using just these definitions, it is easy to see several ways in which tools and technology might be related. Given the first dictionary definition of tool, as a hand held device that facilitates the accomplishment of a task, both the manufacture and application of such devices fits with the first two definitions of technology. A similar connection can be made between the second dictionary definition of tool and all three definitions of technology.

Notice the role that specialization plays in the dictionary definitions. A tool can be something that is "necessary in the practice of a vocation or profession," and technological knowledge can be "the specialized aspects of a particular field." Notice also how broad the third dictionary definition of tool is. Given the definition, "a means to an end," virtually anything can be a tool if it is employed in an *instrumental* way, that is, if it is used in a purposeful way in the pursuit of a goal. I want to incorporate both the role of specialization and the role of goal pursuit to extend the definition of tool to include objects that enhance cognitive as well as sensorimotor capacities, and to force the following two-part distinction between tools and technology. Note that this distinction is, like other definitions, a matter of convenience, a way of providing a frame for further discussion:

1. Tools are discrete, identifiable, physical, organizational, or conceptual objects that facilitate the pursuit of the operator's goals by amplifying the operators sensorimotor and/or cognitive capacities; whereas technology involves the application of systems of accumulative knowledge that allows for the accomplishment of tasks that may or may not be related in any direct way to the goals of the individuals employing the technology.

2. Although tool use can require skills and knowledge that may be difficult to acquire, the acquisition and application of both the skills and knowledge are

transparent and open to anyone with sufficient physical and mental aptitude; whereas the development and application of technology involves a division of labor and specialized knowledge (expertise), that in aggregate form is beyond any single person and thus not completely transparent to anyone involved.

Tools, according to this distinction, have been around for as long as there have been humans, but technology is something brand new, appearing only in the last 5 to 10 thousand years. According to this distinction tools and technology are not synonymous: a cell phone is a tool in the same way that a nuclear weapon is a broom. And the difference is not just a matter of size or scope or flexibility or organizational sophistication. We tend to think of things like cell phones as tools because the part that we manipulate is a small, discrete, handheld object, which is entirely consistent with typical definitions of tool. We forget what it is that a cell phone does— and the average cell phone user has very little understanding of how it does it. We don't see the massive array of towers and sophisticated computer relay stations. We don't see the Asian factories, the rare earth refineries, and the complex corporate infrastructure that are every bit as much a part of the tiny gadget we carry in our pocket. Cell phones are not a product of human manufacture; they are not made by people. There is not an individual or group of individuals who could build a cell phone from scratch. In order for cell phones to exist, it is necessary to first have in place a massively complex economic system, and a sophisticated array of refineries and foundries and factories, and an intricate transportation system—an infrastructure composed of innumerable supportive technologies. It is also necessary to have in place a powerful system of coercion and control: a military-supported political power structure to provide the necessary conditions for organizing the labor to build and maintain this infrastructure. A cell phone does not count as a tool according to either part of the two part distinction above. Cell phone communication is not accomplished by a discrete, identifiable physical object; its operation is a product of the application of highly specialized knowledge, the whole of which is not knowable or transparent to any individual person on earth; and although we can use cell phones to amplify our sensorimotor capacities in the pursuit of goals, more often than not the goals we pursue are not of our free choosing but are forced upon us by the demands of life embedded in techno-industrial system that incorporates cell phone communication.

Nicholas Carr (2010) partitions technology into four categories according to the ways that our inborn capacities are expanded and amplified. The first category is technology that increases our physical ability to operate on the world, and includes a range of devices from the simple sewing needle to a fighter jet. The second category is technology that increases our sensory capacities: the microscope, the Geiger counter, etc. The third category includes technologies that allow us to align nature with our goals and desires. He includes birth control, genetically modified food, and the ability to dam rivers in this category. The final category includes technology that extends our mental powers: the map, the clock, mathematical systems, and the internet fall in this category; presumably so would corporations and other hierarchical systems of administration and government. Although Carr's taxonomic scheme is informative with respect to the range of ways technology can be incorporated into our lives, his partitioning is not associated with any meaningful psychological distinctions. A fighter jet is surely a different kind of thing than a sewing needle from a psychological perspective. Psychologically, a clock and a Geiger counter seem more closely related than do a clock and a multinational corporation. But more importantly, Carr's categories lump tools and technology together.

Although only slightly more psychologically informative, Langdon Winner (1977) provides a far more lucid four-part partitioning of technology. All technology, according to Winner, involves the intentional creation, imposition, addition, or application of structure; there are four general categories to which structure can be applied, yielding four subcategories of technology:

1. Apparatus: instruments, machines, appliances, gadgets (physical objects)

2. Techniques: procedures, methods, skills, routines (structured ways of doing things)

3. Organizations: bureaucracies, schools, armies, corporations (technical social arrangements)

4. Networks: telephone systems, the internet, railroads, highways (systems that combine people and apparatus linked across distance)

Only the first two of these subcategories of technology have any potential overlap with tools according to the two-part distinction I presented above. A physical object can be a tool or a facet of technology depending on how it is being used, whose goals are being served, and whether its use is transparent. Combining technique with apparatus yields what are commonly called crafts. A given technique or craft can be embedded in an opaque system of activity in the service of goals that are not those of the person employing the technique, in which case it qualifies as technology, or it can be a personally controlled, transparent method of facilitating personally-chosen ends, in which case it qualifies as a tool. Neither an organization nor network could ever be a tool because they both involve a division of labor and the partitioning of knowledge, and because they are usually not transparent.

The most important psychological distinction between tools and technology has to do with goals. Tools extend your ability to act in pursuit of personally-chosen ends whereas technology conditions the ends to be pursued—and frequently becomes the end itself as human ends are adjusted to match the features of the technological means, a phenomenon Winner (1977) calls *reverse adaptation:*

[T]echnical systems, once built and operating, do not respond positively to human guidance. The goals, purposes, needs, and decisions that are supposed to determine what technologies do are in important instances no longer the true source of their direction. Technical systems become severed from the ends originally set for them and, in effect, reprogram themselves and their environments to suit the special conditions of their own operation. The artificial slave gradually subverts the rule of the master. (p. 227)

It is not possible to establish whether an apparatus or technique is being used as a tool or functioning as a facet of technology without taking into consideration the goals to which its use is being directed. Take something as simple as a wrench. In the hands of a mechanic maintaining an automobile, a wrench is just another facet of industrial technology: it is necessary for the functioning of the machine in the same way that a specific kind of gear or pulley is necessary. Machines need maintenance, and human wrench-handlers are part of the mechanical process. In the hands of a person disabling a bulldozer that is being used to clear a path through old growth forest, the very same wrench is a tool. Its use is directed at freely chosen ends, and the fact that it was designed so that it neatly fits the bolt-heads of the machine is no different from the fact that a specific kind of hunting knife was designed to separate skin and fur from muscle tissue.

Tools can be complex. Tools can involve division of labor in a superficial sense. The use of some tools requires multiple persons, each performing a specific movement or function at a coordinated time (e.g., casting a large fishing net). And tool manufacture frequently occurs in stages that reflect a systematic, hierarchically organized process—a complex technique, a craft. This is true of the construction of even fairly crude stone tools: first you hew the general shape, then you fashion the cutting edge, then you attach the handle. For more complex tools, each stage can be assigned to a different person, depending on artistic aptitude or experience; but even with many sophisticated crafts, the entire process is transparent to each person involved and each step can be reassigned without a substantial loss of integrity to the final result. Technology, however, involves specialization and the unequal distribution of specialized knowledge. The process is not transparent even to those who are in positions of authority over the process.

Once again, that word, *authority*. Authority is an essential component of technology. The very idea of authority has its source in the division of labor and the isolation of specialized knowledge associated with technology. It is informative to note the two (not mutually exclusive) ways that we commonly use the term *authority:* a person who possesses specialized knowledge (an expert) and a person who has power to control our behavior. Primitive societies have wise elders and others who may be in possession of knowledge that is not in general circulation, but they do not have authorities in either of the senses that we have been trained to accept. Without a broad-based division of labor, authority is unnecessary and, frankly, senseless.

More on this later.

What about human language? Is language a tool or a kind of technology? I'm inclined to say that it is really neither. Although it might be tempting to see language as a technology because it is a system of knowledge that involves the accumulation of structured expressive forms, the absence of *specialized* knowledge (the expressions of a language community are theoretically open to anyone in that community) and the presence of surface transparency (although the inner workings and complexities of language are not consciously accessible) makes it more of a tool. But language qualifies as a tool only in the sense that a hand, when considered in an artificially separate way from the rest of the body, can be considered a kind of tool. That is, it does not qualify as a discrete object in the (mental or physical) environment. Language is employed to achieve social ends, and so is used like a tool; and it can be used to construct additional conceptual and communicative tools much like a hand can be used to construct physical tools. But if a tool is something that is used to enhance an organism's innate capacities, then language doesn't qualify because language *is* an innate capacity. Because language is embodied as an inborn human capacity, it is neither tool nor technology. Language is part of the human configuration. Humans engage in language behavior like they engage in tool behavior: frequently, spontaneously, and in a wide variety of contexts.

One final point about the nature of tools and technology: there is a certain level of adaptation that occurs as a function of personal interaction with either. Anyone who has ever developed calluses from using a shovel, rake, or broom understands how a specific tool can leave its mark. I have a permanent bump on the last joint of the middle finger of my writing hand that reflects decades of pushing pens and pencils. Some tools and many forms of technology necessitate less visible but more severe adaptations. Think of the uncountable cognitive adaptations required to interact with modern computer technology. And even the skilled operation of a primitive stone knife or axe involves permanent neuromuscular adaptation on the part of the operator. My point is that tool use changes the person using the tool. In cases of simple tool use, the changes are usually benign and perhaps even beneficial (e.g., by increasing muscular strength or overall coordination). Accommodating technology, however, can be debilitating, and frequently involves the permanent restructuring of previous ways of living.

Mental Tools

Tools, according to the view I am promoting, can be abstract, conceptual devices as well as physical artifacts. A concept can be a tool of thought, extending and enhancing the effects of a person's cognitive actions in the same way that a physical utensil can extend and enhance the effects of his or her muscular hand movements. A lot has been made of the human capacity for abstraction and symbolic representation. For some, symbolic thought comes closest of any of our latent abilities to being a defining feature of our species. The first clear evidence of mental representation, the harbinger of symbolic capacity, emerges in infancy with the onset of a conceptual ability called *object permanence.* Object permanence is the ability to understand that objects—such as other people—continue to exist when they are no longer perceptually present. In the classic object permanence test, an infant is presented with an interesting toy, and then the toy is hidden behind an opaque barrier or covered with a blanket. Continued interest in the location where the toy was obscured is considered to be evidence that the child knows the toy is still there. Before the onset of object permanence, the child behaves as if the toy is no longer present. Out of sight, out of mind—literally. The lack of object permanence explains the delight young infants take in the game of peek-a-boo. When you place the blanket over your head, you actually disappear from the universe from the pre-symbolic child's point of view. When you suddenly whip the blanket down and say "peek-a-boo," you magically materialized out of nothingness, which is a fascinating trick. Once object permanence develops, the game of peek-a-boo loses much of its surprise value. Object permanence is thought to emerge at about 8 months (Piaget, 1954), although some studies find evidence for rudimentary awareness that objects continue to exist when temporarily obscured in children as young as 3½-months of age (Baillargeon, 1987).

Various products of symbolic thought: abstract representations, schemas, mental models, stereotypes, can be thought of as mental tools, instruments for modifying the cognitive landscape much like concrete tools are used to modify features of the physical environment. The evolution of symbolic thought may be in fact closely yoked to the evolution of physical tool use. At least one theory posits a direct causal connection between tool manufacture and the development of brain areas supporting symbolic thought (Wilkins & Wakefield, 1996): specifically, natural selection for the manual dexterity required for physical tool manufacture led to an increase in the size and complexity of regions of the motor and somatosensory cortex associated with the hands and fingers (the principle of proper mass applies here); the increase in brain tissue in these areas in turn forced areas of association cortex involved in visual, acoustic, and somatosensory processing into closer physical proximity (a region known as the parietal-occipital-temporal junction), which eventually led to the emergence of an area of cortex that received information from three different sensory modalities but was itself functionally independent of any particular modality—and thus provided the neurological underpinnings for abstract representation and symbolic thought.

To focus on the general ability to form symbols or the capacity for abstraction is to somewhat miss the point, however. As generally useful as it is, the capacity for symbolic thought did not evolve, out of the blue, as a general purpose cognitive mechanism. Instead, it likely evolved in conjunction with at least three conceptual abilities that each separately provided a potent fitness advantage. Symbolic thought is a requisite capacity for language. It also serves as a precondition for a theory of mind, and so plays a vital role in cooperative human society. In addition, symbolic thought also provides the undergirding for a general purpose conceptual tool that serves as our primary mode of explanation and comprehension: metaphor. Metaphor allows us to transfer what we learn in one context and apply it to another, superficially unrelated context. It is hard to overstate the power of metaphor as a comprehension tool. It is also hard to exaggerate the role that analogical thought in general, and the use of metaphor in particular, plays in framing our experience (Indurkhya, 2007; Lau & Schlesinger, 2005; Robins & Mayer, 2000; Zualkernan & Johnson, 1992). Landau, Meier, and Keefer (2010), for example, provide extensive evidence that conceptual metaphor plays a fundamental role in social cognition, shaping our thoughts and attitudes concerning events and relationships in our social world.

Metaphor frames how we think, and thus conditions our experiences to follow patterns associated with the specific set of metaphors that are active at the time. Consider how we experience time in a capitalist-industrial society. Lakoff and Johnson (1980) show how the primary metaphor that conditions our experience of time is *time is a resource*. Time is a substance that can be quantified and assigned a per unit value, is directed toward a purposeful end, and is progressively consumed or used up in the service of that end. The *time is a resource* metaphor is inherited from the primary metaphor we use for labor, *labor is a resource,* and the use of time as a means of quantifying labor. Lakoff and Johnson describe how both *time is a resource* and *labor is a resource* metaphors cause us to focus on specific aspects of labor and time that are important in industrial society while at the same time obscuring or deemphasizing other aspects. Potential ways of conceiving time and labor that do not correspond to the resource metaphor, that work can be play, for example, remain hidden by our resource metaphors. And there are other ramifications as well. For example:

> The quantification of labor in terms of time, together with viewing time as serving a purposeful end, induces a notion of *leisure time,* which is parallel to the concept *labor time.* In a society like ours, where inactivity is not considered a purposeful end, a whole industry devoted to leisure activity has evolved. As a result, *leisure time* becomes a *resource* too—to be spent productively, used wisely, saved up, budgeted, wasted, lost, etc. What is hidden by the *resource* metaphors for labor and time is the way our concepts of *labor* and *time* affect our concept of leisure, turning it into something remarkably like *labor.* (Lakoff & Johnson, 1980, p. 67)

The natural world once served as the source of all metaphor. Extensive knowledge of wildlife made animals a powerful metaphoric tool during the Paleolithic. Foraging peoples are keen observers of animal behavior, and Paul Shepard (1982) talks about the ubiquitous use of animals as metaphor in contemporary and ancestral foraging societies. Animals provide a powerful tool for comprehending various aspects of the self and the shifting and interpenetrating nature of relationships between the self and the world. We no longer have animals in our daily experience in anything even remotely resembling the way that our Pleistocene ancestors did. We have pets: dogs and cats, birds and rodents in cages, and fish in glass tanks. And for a minority of people living in rural areas, there is still some degree of exposure to farm animals. But domestic animals are unlike animals in the wild in a number of essential ways. Animal domestication changes the fundamental nature of the relationship and thus the nature of potential metaphors. Consider the few remaining ways that we still use animals as metaphor, and the difference between how we apply wild animals versus domestic animals when referring to human personality characteristics. Wild animals are frequently associated with positive characteristics: clever as a fox, graceful as a deer, wise as an owl, eagle-eyed; whereas domestic animals are frequently used as invective: dumb as a cow, dirty as a pig, stubborn as a mule, chicken (a coward).

Industrial civilization brings with it a host of potential metaphors even as it shuts off access to others. But the metaphors of civilization tend to be superficial, lifeless, mechanical, and inert—in other words, they make for bad metaphor; and so we continue to draw from the natural world. Consider some of the metaphors commonly applied in the financial and economic arena. Economies can be *bulls* or *bears*—animals again, and notice the antagonism between domestic and wild animal. Investments can *grow.* Financial markets can *mature.* There can be *bubbles* and *dips.* Assets can *liquefy* or *freeze.* The problem is that our actual experience of the natural world has become so impoverished that our nature metaphors have lost much of their potential potency. As an aside, it is interesting to note that, unlike symbolic thought, which begins to emerge early in infancy, the capacity for metaphor doesn't begin to bloom until just before puberty. Children prior to the onset of adolescence are notoriously concrete, literal thinkers. If it is true that modern civilization promotes adolescent, and even infantile, levels of immaturity in adults (e.g., Barber, 2007), then we should expect a general decrease in the use of creative metaphor, the fossilization into idiom of commonly used metaphoric expressions, and an increase in literal interpretations of the meaning of events and situations.

Implications, ramifications, and entailments of the capacity for symbolic thought are far too extensive to enumerate here. However, there is one potentially problematic side effect that I want to highlight: the tendency toward reification. Reification involves treating abstract symbolic constructions as if they had an independent existence as material entities in the world. The field of psychology includes a large number of easily reified concepts. My personal favorite example is the idea of *self-esteem*. Self-esteem is a perennially popular research topic. Entering self-esteem as a search term in a popular (non-comprehensive) database for psychology research yielded titles for almost 1700 published studies. Self-esteem is measured and manipulated and serves as the target for intervention strategies as if it was something like muscular strength or insulin tolerance. But it is not like muscular strength and insulin tolerance. Both muscular strength and insulin tolerance are concepts tied directly to concrete physical features of reality. Self-esteem is purely an idea. It serves a conceptual organization function. As a short-hand way of describing a broad cluster of generic characteristics of a person's behavior, self-esteem is a very useful idea. But it is just an idea. There are in theory an infinite number of other possible conceptual schemes that could be applied to organize the phenomena covered by the idea of self-esteem. By contrast, there are only so many valid ways of thinking about muscular strength or insulin tolerance. Other frequently reified concepts from psychology include: intelligence, depression, motivation, cognition, personality, the unconscious, the self, emotions, and mental disorder (Levy, 1997).

Failing to distinguish between an abstract idea and physical reality is a mistake in reasoning Whitehead (1929) called the *fallacy of misplaced concreteness,* a fallacy that "consists in neglecting the degree of abstraction involved when an actual entity is considered merely so far as it exemplifies certain categories of thought" (pp. 7-8). Gilbert Ryle (1949) called this fallacy a *category mistake.* Ryle gave an example similar to the following. You take your elderly grandfather, who never went to college, on a tour of the local university. At the end of the day, after you have toured the entire campus, after you have shown him the libraries, the dormitories, the various departments and offices and classrooms, your grandfather asks when you plan to show him the *university.* Grandpa has committed a simple taxonomic error. He assumed that the university was a singular physical entity, perhaps a kind of building. A university is not just collection of buildings. The idea of university refers to a complex dynamic network of activities, roles, goals, processes, and the physical places within which these things are made manifest. University is way of simplifying reality, a conceptual abstraction, a tool of thought. There is no singular physical entity in the world that corresponds directly to our concept of university.

I want to be clear that reification is not necessarily a bad thing. Sometimes it is quite useful to act in terms of an abstraction as if the target of your action was a material entity in the world. Other times, the failure to distinguish an abstract idea from physical reality can be deadly. And when it comes to comprehension, reified abstractions can obscure as well as facilitate clear understanding. David Hume (1975/1748) pointed out that there are two ways of talking about truth. There are things that are true because we define them as such, for example, that a bachelor is an unmarried male. And there are other things that are true as a function of empirical examination or experience, for example, that water boils at 100 degrees centigrade. Hume's fork, as this distinction is called, calls attention to the fact that some of our ideas are more arbitrary than are others. Although it might be argued that all ideas are at their root social constructions, some of our ideas are linked more closely to physical reality whereas others are designed more for achieving social ends. The reification of the latter can be used to serve political agendas, as when the protection of *liberty, freedom,* or *democracy* are presented as justification for war.

What about civilization? Aren't we talking about an abstraction here? Yes, *civilization,* like *university,* is a shorthand way of referring to a complex aggregate of actual entities and activities. But the claim that, for example, "civilization strips us of our humanity," falls on the empirical tine of Hume's fork: it is a matter of demonstrable fact that the aggregate of entities and activities subsumed under the label *civilization* are in fact dehumanizing. And *megamachine* refers to the actual confluence of technologies, of apparatus, techniques, organizations, and networks, into a coherent, structured—if physically amorphous—whole. *Megamachine* is an abstraction, but it is derived from the particulars of actual circumstances, from empirically-verifiable conditions.

As with metaphor, reification is a powerful potential tool for social control that can be employed in the conversion of free and potentially fierce human beings into meek and servile children. Our capacity for abstraction and symbolic thought allows us to step easily into the role of "servomechanism," and our tendency to reify obscures the contrived nature of our situation. We actually *become* a consumer, employee, homeowner, citizen, as if these roles were something other than conceptual shackles, as if they were obligatory requirements of the universe, as if the machine of civilization was part of the air we breathe. And we become cogs and gears and sprockets grinding against each other, locking each other firmly into place in a circular self-annihilating dance.

We are elephants caught in the diaphanous fibers of a butterfly net.

Humans as Tools

Servomechanisms

One of the most prevalent myths of technology is that its purposes are a reflection of our own: "Every technology is an expression of human will. Through our tools, we seek to expand our power and control over our circumstances—over nature, over time and distance, over one another" (Carr, 2010, p. 44). The truth is exactly the converse: every technological innovation is a further truncation and distortion of human will. "To be commanded, technology must first be obeyed" (Winner, 1977, p. 262). Modern technology increasingly divests us of our individual power and provides additional tendrils of manipulation and control over our personal circumstances. Air travel provides a paradigmatic example of a technology that severely curtails personal freedom and control even as it provides a compelling illusion of power. Let's leave aside for now the obvious freedom-obliterating potential of the military use of air travel, and limit our discussion to the effects of commercial passenger air travel on personal freedom.   In order to travel by air, you need to accommodate the airline's schedule: you are not free to choose when you leave or arrive or, necessarily, your specific destination, you go when and where the airline chooses and can be rerouted in transit without your consent; you have to travel to and from the airport, and are allowed to occupy only certain specific locations within the airport; you are required to submit to severe restrictions on bodily movement: even before you get inside the plane where you are seatbelted into place, you are forced to stand in lines cordoned off like cattle being funneled into the slaughterhouse; and, of course, you are made to submit to invasive security procedures that force you to relinquish all control over the privacy of your physical body and of your personal possessions. And let's not forget the overt monetary costs that are imposed by profit-seeking airline corporations that also, by the way, receive sizeable government subsidies—so you have to pay even if you don't fly, or the immeasurable hidden costs— externalities—of air pollution, noise pollution, resource depletion, habitat destruction, and the general stress and anxiety that life in a jet-setting society kindles. The invention of air travel, more than any other technology prior to the internet, is most directly responsible for the global homogenization of culture and the viral spread of consumerism. Every airplane that takes to the sky reduces the quality of life on the planet and erodes our future freedom some small but irreparable amount. We will return to a more psychologically-relevant and perhaps more ominous example when we look at the adverse effects of technologically-mediated communication below. For now, the main point is that, despite popular notions to the contrary, it is not *our* technology; we—you and I—don't control technology, it clearly controls us. When it comes to the relationship between technology and personal freedom:

> While any new technical device may increase the range of human freedom, it does so only if the human beneficiaries are at liberty to accept it, to modify it, or to reject it: to use it where and when it suits their own purpose, in quantities that conform to those purposes. (Mumford, 1970, p. 185)

I am hard pressed to find a single example of a modern technology that meets these criteria.

According to an *instrumentalist* perspective on technology, technology is neutral and entirely under the conscious control of the people who use it, in other words, technology is just another of our tools; whereas according to a *determinist* view, technology is an irresistible, autonomous, and self-perpetuating force that lies entirely outside anyone's control (Carr, 2010). I suspect that the controversy between these two incommensurate views can be linked to the failure to make a clear distinction between tools and technology, as the distinction I made above seems to largely resolve the apparent conflict: tools, although not entirely neutral in terms of their effects on the user (e.g., calluses, muscle development, eye-hand coordination changes, etc.) are nonetheless largely under the conscious direction and control of the persons operating them. Technology, because of the requisite authority-driven hierarchical division of labor and the role of expertise, tends toward opacity of function and ambiguity with respect to the locus of control. The technological process depends on the coordination of groups of people whose members, as individuals, have limited access to the process itself. Who is in control? Even those responsible for coordinating a major component of the process—say, production or manufacture—cannot be said to *control* the process; much like a musical conductor can *orchestrate*, but does not control the ultimate musical result, which depends on the appropriate action and timing of each individual (expert) musician. Technology is anything but neutral; it serves as a powerful organizing force shaping the thought and behavior of everyone involved. The entire physical infrastructure of any modern American city along with much of the daily activities of its inhabitants is designed around automobile technology, for example. And there is good evidence that interaction with internet technology permanently alters habitual modes of thinking (Carr, 2010). Additionally, specific technologies can have wide ranging consequences for those not directly involved with the technology itself. Consider acid rain caused by emissions from coal power plants, or mercury-laced fish, or nuclear waste, or deforestation and subsequent soil erosion, or, well, you get the point.

The megamachine of civilization is the consummate technology.

The civilizing process is largely a matter of instrumentalizing our human nature, a process where our inborn proclivities are harnessed, distorted, and entrained to the movement of the machine.

Consider, for example, the encounter of an individual with his work "role" in an organization. The whole person in its rich complexity of talents, needs, interests, and commitments is of no use in the performance of the role. Instead, only certain select traits, often created by the role itself and unknown in the life of the individual previously, are demanded by the organization. (Winner, 1977, p. 211)

The denizens of civilization have been conditioned to view each other in instrumental terms. Research has shown that our closeness to significant others increases with their instrumentality, that is, with the extent to which they are important for the achievement of personal goals, and that once a goal has been achieved (or fails to be achieved), our preference for these "instrumental others" diminishes so that we are able to draw closer to other people who are instrumental for alternative goals (Fitzsimons & Fishbach, 2010). Treating others as means to an end is a form of objectification, and the extent to which others are objectified, subjugated, and treated as means to ends is a function of hierarchical power relationships. Social psychologists have found an empirically robust response to social power is that it "involves approaching social targets more when they are useful in terms of an active goal, regardless of the value of their other human qualities" (Gruenfeld, Inesi, Magee, & Galinsky, 2008, p. 123). For those in positions of power, subordinates are seen as tools and as little else: "For power holders, the world is viewed through an instrumental lens, and approach is directed toward those individuals who populate the useful parts of the landscape" (Gruenfeld, Inesi, Magee, & Galinsky, 2008, p. 125). And there is evidence that when relationships are task-oriented and framed in terms of specific goals, people in subordinate positions expect—and even prefer—a clearly defined power differential (Tiedens, Unzueta, & Young, 2007). One of the psychological effects of hierarchical power relationships is acquiescence to our own instrumentalization.

The instrumental nature of our interpersonal interactions has become a ubiquitous feature of consumer society. We have become little more than mechanical means to each other's machine-adapted ends. We are each other's tools, and our interpersonal behavior is rapidly becoming indistinguishable from that of actual machines. Alan Turing, a famous pioneer of digital computing, devised a thought experiment that has been considered a litmus test for artificial intelligence. The experiment goes something like this. You are presented with a computer interface of some sort, a keyboard and screen, for example, asked to engage in a conversation with someone in another room, and told that the person you are conversing with could be either a real person or a computer. Your task is to determine which. If you are in fact interacting with a computer but can't tell whether you are interacting with a machine or a real person, then the computer has passed the "Turing test." I suggest trying the following "reverse Turing test" the next time you are in the local supermarket checkout (assuming that you can still find a checkout station that is crewed by an actual human). Pay close attention to the nature of the interaction between yourself and the checker. Even allowing for quasi-human sounding banter ("My brother tried this spice in his meatloaf last week"), can you distinguish any feature of the interaction that might not be equally effected by an appropriately programmed mechanical device? Is there any substantive difference between your experience with a live checker and your experience with an automated self-checkout station? And how has the incorporation of computer and optical scanner technology over the last few decades changed the experience from the checker's standpoint?

> What merit is there in an over-developed technology which isolates the whole man [sic] from the work-process, reducing him to a cunning hand, a load-bearing back, or a magnifying eye, and then finally excluding him altogether from the process unless he is one of the experts who designs and assembles or programs the automatic machine? What meaning has a man's life as a worker if he ends up a cheap servo-mechanism, trained solely to report defects or correct failures in a mechanism otherwise superior to him? If the first step in mechanization five thousand years ago was to reduce the worker to a docile and obedient drudge, the final stage automation promises today is to create a self-sufficient mechanical electronic complex that has no need even

for such servile nonentities. (Mumford, 1970, p.179)

Mediated Communication

Perhaps nowhere is the impact of modern industrial technology on our personal lives more apparent than with our communication technology. As a species, our communicative capacities have no rival. In addition to spoken language, the sine qua non of human culture, we possess highly articulated nonverbal receptive and expressive abilities. We can send and receive complex, subtle, multimodal messages, the contents of which are open-ended, unrestrained by time or space or conceptual category. Our elaborate capacities of communicative interaction have been fine-tuned by millions of years of evolution as a social animal embedded in small-group, face-to-face communities. Once we step away from real-time, co-present interaction, our communicative potential is greatly diminished, the content of our messages is restricted, and our interpretive abilities are impoverished.

Real-time, co-present communication is not an efficient means of messaging when the population exceeds the limits of a small village; and it is entirely unnecessary when human relationships have been mechanized and the messages are merely procedural instructions, or, perhaps more frequently, a mildly entertaining mode of distraction. The differences between an exchange of text messages and a face-to-face conversation are massive and qualitative. But because our relationships have been rendered two-dimensional, we hardly notice the difference. Because the majority of our important interactions (with other people and with the rest of the natural world) are mediated by the machine of civilization, we don't miss what is missing.

Physical signs, marks carved or painted on rocks and trees were among the very first forms of mediated communication. Although physical signs remove the temporal limitations of spoken language, this one advantage does not begin to compensate for what is lost in the process: it is easy to see the communicative limitations of a petroglyph. Writing retains the time-travelling advantage of the petroglyph while extending the potential complexity of the message by several orders of magnitude, and it has the additional advantage of being portable. The advantages associated with written language are so great that the fact that it is still a dramatically impoverished form of communication relative to the co-present face-to-face spoken word goes largely unrecognized. Writing involves the loss of immediacy, real-time disambiguation and feedback, spatial and temporal context sensitivity, and intimacy, among other things. But even more profoundly, the incorporation of writing technology into a society changes the nature of the society in dramatic ways. Perhaps most notably, like all technology, it can quickly lead to a dependence relationship. Where once it was necessary to exercise the power of memory and employ various mnemonic tools such as song, rhythm, and rhyme to retain important details, with writing, these details can be physically preserved. The need to trust the integrity of the speaker—and perhaps the importance of honesty as a social value—diminishes as the physical document becomes the vessel of truth. The addition of writing also creates a class distinction between the literate and the non-literate within a society. Access to power increases with literacy, a fact that is reflected in the strong correlation found between literacy and health and longevity (Marks, 2007). Writing has been used as a tool of subordination, subjugation and conquest since its inception, beginning at least as far back as the Code of Hammurabi. The ways that written text in the form of poorly understood treaties was (and is still being) used in the genocidal conquest of indigenous North Americans is well documented.

The term "preliterate" is a loaded term. It assumes that the lack of written language is evidence of cultural inferiority. Cultures without written language are considered *primitive,* in the progressive-delusion sense that they are not as far along on the progressive timeline of cultural evolution. However, the presence or absence of literacy in a society has nothing to do with intellectual complexity or sophistication. Cross-culturally, there is no relationship between literacy and intelligence (Duffy, 2000). Non literate Australian Aborigines possess probably the most elaborate and sophisticated cosmology of any people, for example (Shepard (1998), a cosmological understanding that is more nuanced than the modern scientific perspective and far more coherent than the written-text-based medieval Christian view. Whether or not a traditional non literate culture adopts the technology of written language has to do with its exposure to, and relationships with, cultures of power, and nothing to do with so-called cultural evolution, a rhetorical device contrived by dominant cultures to justify the continued exercise of their dominance (Duffy, 2000).

Historically, within Western culture written language was restricted to a minority of literate specialists until the invention of moveable type. With the printing press, the written word became more generally available to the population, and the ability to read, while still limited to the upper classes, was no longer restricted to a minority of specially trained experts. Writing (and reading) changes how we think about things, enhancing and emphasizing a linearity that is already latent in spoken communication. Lewis Mumford (1934) suggests that writing changed the fabric of thought itself, and that the "lapse of time between expression and reception had something of the effect that the arrest of action produced in making thought itself possible" (p. 239). Written language allows for the formation of deeper, more abstract, and more reflective connections, and makes more "pregnant the intercourse of men" (Mumford's words). Nicholas Carr (2010) argues that the hypertext of the internet, because of its ubiquitous choice points, its ever-branching links, engenders a change in the fabric of thought on par with, and perhaps more extensive, than that caused by exposure to linear written text.

The late nineteenth and twentieth century saw the emergence of electronically mediated communication, first with the telegraph, and then the telephone and radio, and finally with the digital computer and cellular telephone. Electronic mediation retains the distance-conquering features of written text, while re-introducing the immediacy of spoken face-to-face communication. The net result has not necessarily been positive, however. Mumford, speaking prophetically in

1934 cautioned against assuming future innovation in technology would automatically lead to an improved communicative situation:

> What will be the outcome? Obviously a widened range of intercourse: more numerous contacts: more numerous demands on attention and time. But, unfortunately, the possibility of this type of intercourse on a worldwide basis does not mean a less trivial or less parochial personality. For over and against the convenience of instantaneous communication is the fact that the great economical abstraction of writing, reading, and drawing, the media of reflective thought and deliberative action will be weakened. [...] That the breadth and too-frequent repetition of personal intercourse may be socially inefficient is already plain through the abuse of the telephone: a dozen five minute conversations can frequently be reduced in essentials to a dozen notes whose reading, writing, and answering takes less time and nervous energy than the more personal calls. With the telephone the flow of interest and attention, instead of being self-directed, is at the mercy of any strange person who seeks to divert it to his own purposes (p. 240).

One might easily develop a critique of the putative benefits of email or cell phone text messaging similar to the one Mumford leveled against the telephone of his time.

Our ability to engage in "pregnant intercourse" with each other electronically has increased dramatically the last few decades. Paradoxically, so has our social distance from each other. Our enhanced ability to connect with others has not made us more connected. Social isolation actually increased in the 19 years from 1985 to 2004, as first the personal computer, and then the internet and cell phones, became an integral feature of our communicative landscape. The average number of confidants in a person's intimate social network decreased from almost three in 1985 to two in 2004, and the number of people reporting that there is not a single person with whom they can confide important matters tripled during that same time period (McPherson, Smith-Lovin, & Brashears, 2006). In addition, research directed specifically at the social and emotional impact of the use of online social media has failed to find any real-life benefits (Pollet, Roberts, & Dunbar, 2011), and several potential threats to personal wellbeing (Kalpidou, Costin, & Morris, 2011). Clearly our sophisticated digital communication technology is not improving our social relationships.

And there is substantial evidence that electronically-mediated communication impacts more than just our social lives. Psychologically significant differences between face-to-face and technologically-mediated communication have been known for years. An early review of the empirical research in this area found little difference between face-to-face and mediated communication when participants are given a cooperative task, however, when the task involves conflicting goals several differences emerge, with mediated communication leading to (1) a reduction in the likelihood that a conclusion will be reached based on the merits of the case, (2) better success at arguing for a position the participant didn't believe, (3) less attention to emotion-related content, and (4) lower evaluations and worse impressions of the other person (Williams, 1977). A slightly more recent study (Doherty-Sneddon, et al., 1997) found that the nature of dialog changes in cooperative tasks as a function of mediation, with mediated communication involving extra utterances directed at eliciting feedback that would normally have been provided by nonverbal cues. This result is perhaps not surprising. Interestingly, this latter study also found that the inclusion of visual nonverbal cues mediated through high quality video did not appear to change things much, and did not produce benefits comparable to unmediated face-to-face interaction. This finding is reminiscent of the failure of a high resolution real-time video nature scene to induce the calming effect produced by viewing the same scene through a window, discussed earlier: the electronic flattening of the visual world appears to strip out something essential. The lack of nonverbal cues and the spatial and temporal isolation involved with email and text message communication can reduce the "reality-check" we get from real-time exposure to other people's perspectives, and can lead to overconfidence in our ability to accurately interpret the meaning and intent of the other person's message. We have difficulty getting outside of our own perspective when we assess the perspective of the other person, but that fact is obscured from us because of an egocentrism that is generated by the isolating context (Kruger, Epley, Parker, & Ng, 2005). The risk of misinterpreting sarcasm is extremely high, for example.

The potential psychological impact of the internet is still largely uncharted territory. But there is reason to suspect that there may be some wide-ranging adverse effects of chronic internet usage. "The net's interactivity gives us powerful new tools for finding information, expressing ourselves, and conversing with others. It also turns us into lab rats constantly pressing levers to get tiny pellets of social nourishment" (Carr, 2010, p. 117). Internet addiction, while not yet an official diagnostic category, is generally accepted as a psychiatric reality (Thorens, et al., 2009), and appears to share many of the same behavioral characteristics with drug addiction, including increased impulsivity and a reduction in the ability to delay gratification because of a tendency to discount the value of delayed rewards (Saville, Gisbert, Kopp, & Telesco, 2010). Online social interaction has been found to be positively correlated with compulsive internet use and depression in adolescents (van den Eijnden, Meerkerk, Vermulst, Spijkerman, & Engels, 2008), and teen suicide resulting from cyberbullying is making regular headlines.

Mediated communication changes how we treat each other. The distancing and depersonalization that is associated with electronically-mediated group communication, for example, has been found to cause a smoothing-out of perceived individual differences, an increased focus on group membership, and an increased use of stereotypes (Postmes, Spears, & Lea, 2002). Mediation changes how we think about ourselves as well. Incorporation of mechanical automation into a complex task can have an impact on self-confidence as compared to incorporating the assistance of actual persons: when something goes wrong with an automated process, self-confidence of the person in charge tends to decrease, but when the same error occurs when real people have been given control of the process, self-confidence of the person in charge is unaffected (Lewandowsky, Mundy, & Tan, 2000).

Our communication technology isolates and insulates us from each other and mechanically conditions both our interpersonal interactions and our introspective awareness. Electronic mediation further blurs the line between human and machine, and increases the extent to which we treat each other as objects, as standardized units in a planet-wide production process rather than as unique individuals, as instruments rather than human beings.

The Birth of the Machine

The Domestication Frame

The Neolithic marks the beginnings of large scale domestication, what is typically referred to as the agricultural revolution. It was not really a revolution in that it occurred over an extended period of time (several thousand years) and in a mosaic piecemeal fashion, both in terms of the adoption of specific agrarian practices and in terms of specific groups of people who practiced them. Foraging lifestyles continue today, and represented the dominant lifestyle on the planet until relatively recently. The agricultural revolution was a true revolution, however, in terms of its consequences for the humans who adopted domestication-based life-ways, and for the rest of the natural world. The transition from nomadic and seminomadic hunting and gathering to sedentary agriculture is the most significant chapter in the chronicle of the human species. But it is clearly not a story of unmitigated success. Jared Diamond, who acknowledges somewhat the self-negating double-edge of technological "progress," has called domestication the biggest mistake humans ever made.

> That transition from hunting and gathering to agriculture is generally considered a decisive step in our progress, when we at last acquired the stable food supply and leisure time prerequisite to the great accomplishments of modern civilization. In fact, careful examination of that transition suggests another conclusion: for most people the transition brought infectious disease, malnutrition, and a shorter life span. For human society in general it worsened the relative lot of women and introduced class-based inequality. More than any other milestone along the path from chimpanzeehood to humanity, agriculture inextricably combines causes of our rise and our fall. (Diamond, 1992, p. 139)

The agricultural revolution had profoundly negative consequences for human physical, psychological, and social wellbeing, as well as a wide-ranging negative impact on the planet.

For humans, malnutrition and the emergence of infectious disease are the most salient physiological results of an agrarian lifestyle. A large variety of foodstuffs and the inclusion of a substantial amount of meat make malnutrition an unlikely problem for hunter gatherers, even during times of relative food scarcity. Once the diet is based on a few select mono-cropped grains supplemented by milk and meat from nutritionally-inferior domesticated animals, the stage is set for nutritional deficit. As a result, humans are not as tall or broad in stature today as they were 25,000 years ago; and the mean age of death is lower today as well (Sheppard, 1998). In addition, both the sedentsim and population density associated with agriculture create the preconditions for degenerative and infectious disease. "Among the human diseases directly attributable to our sedentary lives in villages and cities are heart and vascular disorders, diabetes, stroke, emphysema, hypertension, and cirrhoses [sic.] of the liver, which together cause 75 percent of the deaths in the industrial nations" (Sheppard, 1998, p. 99). The diet and activity level of a foraging lifestyle serve as a potent prophylactic against all of these common modern-day afflictions. Nomadic hunter-gatherers are by no means immune to parasitic infection and disease. But the spread of disease is greatly limited by low population density and by a regular change of habitation which reduced exposure to accumulated wastes. Both hunter-gatherers and agriculturalists are susceptible to zoonotic diseases carried by animals, but domestication reduces an animal's natural immunity to disease and infection, creates crowded conditions that support the spread of disease among animal populations, and increases the opportunity for transmission to humans. In addition, permanent dwellings provide a niche for a new kind of disease-carrying animal specialized for symbiotic parasitic cohabitation with humans, the rat being among the most infamous. Plagues and epidemic outbreaks were not a problem in the Pleistocene.

There is a significant psychological dimension to the agricultural revolution as well. A foraging hunter-gatherer lifestyle frames natural systems in terms of symbiosis and interrelationship. Understanding subtle connections among plants, animals, geography, and seasonal climate change is an important requisite of survival. Human agents are intimately bound to these natural systems and contemplate themselves in terms of these systems, drawing easy analogy between themselves and the natural communities around them, using animals, plants, and other natural phenomena as metaphor. The manipulative focus of domestication frames natural systems in antagonistic terms of control and resistance. "Agriculture removed the means by which men [sic.] could contemplate themselves in any other than terms of themselves (or machines). It reflected back upon nature an image of human conflict and competition..." (Shepard, 1982, p. 114). The domestication frame changed our perceived relationship with the natural world, and lies at the heart of our modern-day environmental woes. According to Paul Shepard, with animal domestication we lost contact with an essential component of our human nature, the "otherness within," that part of ourselves that grounds us to the rest of nature:

> The transformation of animals through domestication was the first step in remaking them into subordinate images of ourselves—altering them to fit human modes and purposes. Our perception of not only ourselves but also of the whole of animal life was subverted, for we mistook the purpose of those few domesticates as the purpose of all. Plants never had for us the same heightened symbolic representation of purpose itself. Once we had turned animals into the means of power among ourselves and over the rest of nature, their uses made possible the economy of husbandry that would, with the addition of the agrarian impulse, produce those motives and designs on the earth contrary to respecting it. Animals would become "The Others." Purposes of their own were not

allowable, not even comprehensible. (1998, p. 128)

Domestication had a profound impact on human psychological development. Development—both physiological and psychological—is organized around a series of stages and punctuated by critical periods, windows of time in which the development and functional integration of specific systems are dependent upon external input of a designated type and quality. If the necessary environmental input for a given system is absent or of a sufficiently reduced quality, the system does not mature appropriately. This can have a snowball effect because the future development of other systems is almost always critically dependent on the successful maturation of previously developed systems. The change in focus toward the natural world along with the emergence of a new kind of social order interfered with epigenetic programs that evolved to anticipate the environmental input associated with a foraging lifestyle. The result was arrested development and a culture-wide immaturity:

> Politically, agriculture required a society composed of members with the acumen of children. Empirically, it set about amputating and replacing certain signals and experiences central to early epigenesis. Agriculture not only infantilized animals by domestication, but exploited the infantile human traits of normal individual neoteny. The obedience demanded by the organization necessary for anything larger than the earliest village life, associated with the rise of a military caste, is essentially juvenile and submissive... (Shepard, 1982, pp. 113-114)

Along with general psychological changes, the domestication frame led to changes in spiritual beliefs and religious practices. Hunter-gatherer religious sentiments are based on a diffuse and embedded spirituality in which humans and other animals participate equally. Contact with the spirit world is continual and unmediated, and it is shared equally, not only by all members of the culture, but by all organisms. There is no need for intermediaries or religious specialists. The domestication frame changes the orientation toward the natural world from a lateral, shared participation to a dominance hierarchy with humans above and the products of human manipulation and control below. Animals become subordinate, dumb inferiors, soulless creatures. When people step back to contemplate their own position in the cosmos and bring this hierarchical framework with them, the presence of a superordinate deity or deities becomes an unavoidable inference. And unlike the foraging situation where contact with the spiritual other was direct, equally shared, and unmediated, contact with a higher-order deity requires religious specialization. Domestication brought with it two distinct religious tendencies, two poles on a continuum, one associated with farmers and the other with pastoralists. Farming lends itself to a terrestrial focus and to religious practices associated with the worship of Earth-Mother deities linked to important physical locations, geographic regions, or geological formations (e.g., springs), whereas herding lends itself to a concern with atmospheric whether events and a skyward focus, and to the worship of impulsive and vengeful masculine deities that are portable in that they are not specifically affiliated with any geographic location (Shepard, 1998). The Judeo-Christian-Moslem god obviously derives from the pastoralist archetype. The important point is that, with domestication, deities emerge as distinct targets of worship and ritualized petition—often through the mediation of specialists, the shaman, priest, or priestess—and that the relationship between deity and human supplicant mirrors in nontrivial ways the relationship between humans and the living targets of human domestic control and manipulation.

The transition to agricultural lifestyles was both result and cause of a variety of rather dramatic social changes. The most important of these for our present purposes is that domestication of plants and animals provided a new set of metaphors and a new way of framing social relationships in terms of hierarchical systems of power and control. In foraging cultures and "non-Western, un-industrialized, and largely illiterate (hence nonhistorical) societies, power is plural, societies are egalitarian, and leadership is not monopolized but changing and dispersed" (Shepard, 1998, p. 60). But with the transition to an agrarian lifestyle come the seeds of centralized power. It is impossible to overstate the potential impact centralized power can have on individual freedom. Additionally, centralized power requires the development of structures for exercising that power. As we will see shortly, these structures served as the prototype for all complex mechanical systems—physical and otherwise. But mechanical power is not achieved through organizational structure alone; the gears have to have teeth. So along with the emergence of centralized, hierarchical systems of power came something that was completely absent during the Pleistocene: the manufacture of military weapons.

> Out of the early neolithic complex a different kind of social organization arose: no longer dispersed in small units, but unified in a large one: no longer 'democratic,' that is, based on neighborly intimacy, customary usage, and consent, but authoritarian, centrally directed, under the control of a dominant minority: no longer confined to a limited territory, but deliberately going 'out of bounds' to seize raw materials and enslave helpless men, to exercise control, to extract tribute. This new culture was dedicated, not just to the enhancement of life, but to the expansion of collective power. By perfecting new instruments of coercion, the rulers of this society had, by the Third Millennium B.C., organized industrial and military power on a scale that was never to be surpassed until our own time (Mumford, 1966, p. 164).

The domestication frame, the late Neolithic emergence of centralized power—"the great curse of human history" (Shepard, 1998, p. 156)—and the development of physical tools of coercion and violence were preconditions for the assembly of the first megamachines of civilization in Mesopotamia and Egypt.

The First Megamachine

There is no empirical support for the commonsense model of a linear progressive cultural evolution from foraging bands to tribes to chiefdoms to states (Abrutyn & Lawrence, 2010). Nor is their sufficient data to support an "emergent property" conception whereby states, or civilizations, emerge spontaneously as a natural feature of the complexity associated with population density in a region. The transition to civilization, according to this latter view, is something akin to the transitional states of water. When the temperature gets below 32 degrees Fahrenheit, water molecules spontaneously assume a particular kind of organization, and ice crystals form. Likewise, when a regional human population reaches a certain density, cities coalesce and civilization spontaneously emerges. Humans, however, are not like water molecules. Civilization is not a natural state for humans; it is a hierarchical system of power and control, a machine. The machine of civilization requires the presence of certain conditions including a sedentary (non-nomadic) lifestyle and a minimum population density in order to be applied, a substrate through which its machinations can operate, but it represents the accumulative consequences of specific historical events—each of which may be entirely unique to a given civilization. It is not an emergent property of those historical events any more than the flooding of the Nile is an emergent property of the molecular structure of water or a 67 Chevy Impala is an emergent property of metal.

The specific historical events leading from communal Near East Neolithic farming villages to the first city-anchored agrarian states in Mesopotamia and Egypt are controversial. In some aspects, the transition appears to have been a gradual and continuous, more or less linear process of adaptation to changing social conditions; in others, the transition displayed a punctuated stage-like series of rapid changes followed by periods of little or no change, and resembled the pattern of punctuated equilibrium seen with biological evolution (Abrutyn & Lawrence, 2010). The archaeological evidence suggests that the final shift from Neolithic chiefdom to the first urban-centered agrarian states was rapid and involved dramatic qualitative changes in social circumstances (Abrutyn & Lawrence, 2010). States and chiefdoms are in fact qualitatively different systems, and there are some notable differences between the two having to do with the capacity to exert power and control. Power in a chiefdom is limited and unstable, and its legitimacy is subject to negation by a wide variety of external and internal contingencies. The agrarian state, however, has an enforceable monopoly on the means of violence, and the legitimacy of power is typically supported by divine proclamation. The monopoly on violence and the sanction to use as much violence as is required to maintain control has perhaps reached its apex in our own modern police state. As we will see in the Part 3, the systemic requirement to meet all resistance with an overwhelming force in order to maintain the monopoly on violence is an essential feature of the modern machine.

Chiefdoms are based on kinship and agreement—and held together by the wile and fortune (in both the monetary and serendipity senses of that term) of specific persons. States, as complex systems of power and control, are not dependent on the capacities or kinship ties of any single individual. The sovereign is frequently just a figure head. What was important for the early states was that the sovereign was perceived by the masses to have a divinely ordained reason for occupying that position. For a hierarchical system of power to be effective, the source of power at the top needs to be seen as valid. By assigning the ultimate level of power to a divine source, its validity was unquestionable so long as war and natural disaster didn't intervene and bring divine sanction into question. The hierarchical system of power that came into existence with the first city-states was not just a new kind of social-political system. According to Lewis Mumford, the emergence of the state under divine kingship represents an important culminating step in the assembly of the archetypal machine, the prototype of all mechanical systems to follow. "This extraordinary invention proved in fact to be the earliest working model for all later complex machines, though the emphasis slowly shifted from human operatives to the more reliable mechanical parts" (1966, p. 188). Mumford sometimes referred to this as the *invisible machine* because its various parts were separated in space even when it was in full operation. He sometimes called it the *labor machine* when it was applied to large-scale collective enterprises, building pyramids or irrigation canals for instance. He called it the *military machine* when it was directed toward "acts of collective coercion and destruction" (p. 188). When referring to the vast oppressive networks of domination and control that emerge in our modern technological civilization, he called it the *power complex*. When taken as a whole, when the operation and interactions among all of its political, economic, bureaucratic, and military components are taken together, it is the *megamachine*.

> Now to call these collective entities machines is no idle play on words. If a machine be defined, more or less in accord with the classic definition of Franz Reuleaux, as a combination of resistant parts, each specialized in function, operating under human control, to utilize energy and perform work, then the great labor machine was in every aspect a genuine machine: all the more because its components, although made of human bone, nerve, and muscle, were reduced to their bare mechanical elements and rigidly standardized for the performance of their limited tasks (Mumford, 1966, p. 191).

In *Technics and Human Development,* Mumford outlined the structure and function of the megamachine assembled in ancient Egypt to build the Great Pyramid of Cheops. "This megamachine was composed of a multitude of uniform, specialized, interchangeable but functionally differentiated parts, rigorously marshaled together and coordinated in a process centrally organized and centrally directed: each part behaving as a mechanical component of the mechanized whole" (p. 196). Two devices had to be in place and operational in order for the machine to function: a reliable system for organizing and controlling knowledge, and a system for issuing orders and following them through.

> The first was incorporated in the priesthood, without whose active aid the institution of divine kingship could not have come into existence: the second, in a bureaucracy. Both were hierarchical organizations at whose apex stood the high priest and the king. Without their combined efforts the power complex could not operate effectively. This condition remains true today, though the existence of automated factories and computer-regulated units conceals both the human components and the religious ideology essential even to current automation (Mumford, 1966, p 199).

The great pyramid was constructed through the implementation and coordination of three invisible machines, each of which is in operation today: the military machine, the labor machine, and a communication bureaucracy. The modern versions of these invisible machines are more efficient and effective, and the scope and structure of the global megamachine they comprise is orders of magnitude greater in both size and complexity than the one assembled in ancient Egypt. The industrial revolution was essentially a retooling of the labor machine, which led to subsequent upgrades in both the military and bureaucratic-communication machines, but the fundamental design of the megamachine has not changed since its original inception. Neither has its ultimate purpose: to organize, entrain, and direct human activity.

> [A] close parallel exist[s] between the first authoritarian civilizations in the Near East and our own, though most of our contemporaries still regard modern technics, not only as the highest point in man's [sic] intellectual development, but as an entirely new phenomenon. On the contrary, I found that what economists lately term the Machine Age or the Power Age, had its origin, not in the so-called Industrial Revolution of the eighteenth century, but at

the very outset in the organization of an archetypal machine composed of human parts. (Mumford, 1966, p. 11)

All complex mechanical systems involve a hierarchical structuring of power. The organized distribution of power and the patterns of resistance among the various components are what allow the machine to function. This is a matter of physics in a machine composed of metal parts. Human psychology provides an additional layer of complexity in a machine composed of human parts. Human resistance is less predictable and can take a variety of forms. Despite the high degree of interchangeability, not all humans are equally good candidates for a particular role in the process. "Only those who were sufficiently docile to endure the regimen—or sufficiently infantile to enjoy it—at every stage from command to execution could become efficient units in the human machine" (Mumford, 1966, p. 202). With the first megamachines, those who were not "sufficiently docile" were simply killed; in the modern megamachine, they are pushed to the periphery of society, or, more frequently, flushed into the prison system.

Always reserving the exclusive privilege to engage deadly force, the modern machine has replaced overt force and direct threat in most cases with more subtle means of coercion and conditioning. The tools of domestication have been reworked, refined, and recalibrated for use with humans. Humans, like their domesticated animal equivalents, have been rendered passive and dependent in the process.

Already this mode of conditioning has created a new psychological type: one bearing almost from birth the imprint of megatechnics in all its forms: a type unable to react directly to sights or sounds, to patterns or concrete objects, unable to function in any capacity without anxiety, indeed, unable to feel alive, except without permission or command of the machine and with the aid of extra-organic apparatus that the Machine-God provides. In a multitude of cases, this conditioning has already reached the point of total dependence; and the state of submissive conformity has been hailed, by the more ominous prophets of this regime, as man's ultimate 'liberation.' But liberation from what? Liberation from the conditions under which man has flourished: namely, in an active, give-and-take, mutually rewarding relationship with a varied and responsive 'un-programmed' environment, human and natural—an environment full of difficulties, temptations, hard choices, challenges, lovely surprises, and unexpected rewards (Mumford, 1970, p.284).

Modern social science, psychology in particular, has done much to legitimize the conditioning process and intensify our dependence. Ergonomic factory design, psychotherapy, and psychopharmaceutical regimens serve as anesthesia, and the availability of—and obligatory participation in—continual distraction offered by high-tech entertainment technology coaxes us into deeper and deeper levels of dependency; and we become further and further removed from the conditions in which "flourishing" is even theoretically possible.

And the invisible machine grinds on.

Although historically bumpy, with numerous twists and turns and branching regressions, the psychological trail from domestication to compulsory human subservience in a high-tech planet-wide hierarchical system of power is a smooth unbroken pathway. And our gleaming skyscraper-bursting megalopolises with their ever-expanding fossil carbon incandescence are simply a modern variation on the domestication theme, different in magnitude but not in kind from the earliest walled citadels of Mesopotamia and Egypt. "Why this 'civilized' technical complex should have been regarded as an unqualified triumph, and why the human race has endured it so long, will always be one of the puzzles of history." (Mumford, 1966, p. 213)

Maintaining the flow of power

Power in the megamachine derives in part from the hierarchical nature of the system itself and the way that knowledge is partitioned. Knowledge is power. Lack of essential knowledge is the source of dependency. Knowledge is what separates those in superordinate positions of the hierarchy from their subordinates, and is meted out to subordinates strictly on a need to know basis. If the subordinate possessed all the knowledge that the superordinate possessed, dependency would vanish and there would be no need for the superordinate in the first place. Bureaucracies can expand to Rube-Goldberg levels of complexity as a function of the partitioning and greedy sheltering of minute provinces of knowledge. To maintain the integrity of the system as a whole, knowledge of how the upper-reaches of the system operate must be highly restricted.

> To be effective, this kind of knowledge must remain a secret priestly monopoly. If everyone had equal access to the sources of knowledge and to the system of interpretation, no one would believe in their infallibility, since the errors could then not be concealed [...] Secret knowledge is the key to any system of total control. Until printing was invented, the written word remained a class monopoly. Today the language of higher mathematics plus computerism has restored both the secrecy and the monopoly, with a consequent resumption of total control (Mumford, 1966, p. 199).

One way that knowledge is controlled in our society is through the sanctioning of experts. Consider two salient examples: the sanctioning of medical specialists and the partitioning of academic disciplines into minutely focused and highly guarded fields of expertise. In both cases, dependency on the system is established and maintained by restricting access to knowledge. In the former case, it is illegal for the non-specialist to apply protected information that might accidently leak out into the general knowledge pool. There are severe restrictions on what kinds of chemicals a person is allowed to ingest without the express written permission of a medical specialist: legal sanctions enforced by lengthy prison terms for anyone who self-medicates with controlled pharmaceutical substances. The restrictions on prescription medications are justified in the name of protecting the public. However, an objective comparison of the potential dangers of prescription and nonprescription medication does not warrant any distinction whatever, and evidence that laws concerning the distribution of pharmaceutical substances serve the economic interests of pharmaceutical corporations is legion. The academic world also has legal barriers against the general distribution and application of knowledge: copyright and intellectual property laws limit the free flow of information and help to maintain dependency on specially sanctioned experts in various fields of study. In many cases, general public access to scientific research is not available; access to information is limited to paying subscribers or persons affiliated with a specific scientific, educational, or professional community—people with a "legitimate" right to access.

In addition to the tight control of knowledge, the hierarchical structure of the machine provides more direct ways of channeling power through the system, namely, through cascading layers of dominance and authority. Dominance and obedience to authority are built in to the interpersonal roles we play and dictate the course of our public interaction. We do not necessarily have to be informed explicitly about the particular dominance relationships that are in operation in any given social context. Research has demonstrated that people process information about dominance and submissiveness automatically, utilizing nonverbal cues such as body posture and symbolic indications of power and authority (Moors & De Houwer, 2005). When we engage in public intercourse, we assume that we are embedded in a system of dominance and submission, of authority and expected obedience.

All social primates show some degree of social stratification according to dominance. For humans, nature divides dominance among a variety of context-dependent categories, physical strength, prior experience, personality, and leadership ability, for example. The megamachine partitions dominance additionally according to mechanical role in the hierarchy, sometimes in a way that overlaps with natural categories, as when a person with a personality predisposed to leadership is assigned a dominant role. Frequently, however, the match between natural leanings, assigned role, and context do not comprise a coherent package. The public education system serves an important sorting function in this regard, sifting out people who are competent to serve at various levels in the system. The larger point is that another way that the mismatch between our genetic expectations and civilization manifests is in the lack of consistency between our inborn tendencies regarding dominance and submission and the machine's artificially-imposed hierarchy and demand for obedience to arbitrary authority.

Frequent resistance to authority would appear to be a logical outcome of the mismatch, a result that would appear on the surface to be counter to the interests of the machine. Paradoxically, a certain amount of resistance is necessary in order for the machine to function. Although people automatically perceive dominance relationships, people don't consent to the dominion of authority automatically solely as a function of internalized social norms or desires that operate preconsciously; rather, both compliance and consensus result from an ongoing interaction among "heterogeneous and calculative but mutually susceptible individuals," and resistance to power plays a central role in the formation and maintenance of routinized systems of control (Rafanell & Gorringe, 2010). A certain amount of pushback from below is necessary in order for power from above to get any traction. Thus, to anticipate part of the discussion of resistance in Part 3, taking a fight-fire-with-fire approach to confronting power is likely to backfire—especially when there is a large power disparity to begin with. In our own time, we have seen that active resistance is frequently used to justify a further reduction of freedom and an increase in force from above. And let's not forget that the machine will use overwhelming force to maintain its monopoly on violence. Resistance serves the machine by identifying potentially exploitable weaknesses in the systems of control that need to be fortified. When it comes to resistance from below, to borrow and twist Nietzsche's most misquoted slogan: that which does not kill the machine makes it stronger.

Civilization's social programming makes true resistance an exceedingly rare phenomenon, however. Although we are prepared by our primate evolutionary heritage to navigate dominance relationships, obedience to authority is not part of our evolved psychological tool kit. The need to be obedient has to be impressed upon us; it has to be conditioned.

> The technology of behavior control begins with the subtle family processes involved in "civilizing" infants to be "good," "acceptable" children. The major lesson taught in all traditional school systems is the necessity to obey trivial, irrelevant rules and to observe protocol, while at all times respecting authority because it exists. Such control is perfected by a variety of social institutions that encourage adults to exchange freedom of thought, independence, and individuality for the delusions of social and economic security, collective political strength, and personal approval (Zimbardo, 1974, p. 566).

Note in this context the phrase, "respect for authority." This common parental admonition is an example of how reification can be used to take an existing psychological tendency, in this case a natural sensitivity to dominance relationships among individuals, and yoke it to a mechanical requisite of civilization, in this case, the need to coordinate with an artificial power hierarchy. Authority is not the kind of thing that could possibly be the target of the cluster of psychological postures that are subsumed by the word *respect*. You can respect individual people who hold positions of authority, but the only possible responses to authority itself are obedience and compliance, and their opposites.

In the early 1960s Stanley Milgram, in what has become a

classic set of studies on the nature of obedience, demonstrated the coercive power of authority even in situations where no real authority exists. Milgram (1963) had people who thought they were participating in a learning experiment deliver what they thought were increasing levels of electrical shock to what they thought were fellow participants in the experiment. Milgram designed a simulated shock generator consisting of a large electronic device with 30 toggle switches labeled with voltage levels starting at 30 volts and increasing by 15-volt intervals up to 450 volts. The switches were further labeled in groups such as "slight shock," "moderate shock," and "danger: severe shock." The learning task involved the learner memorizing connections between lengthy lists of word pairs. The "teacher" would read the list, and then test the learner's memory for them. The learner was really a confederate, an actor wired up to the fake shock generator and instructed to fail to learn and pretend to experience shock. The basic procedure involved instructing the teacher to give an electric shock each time the learner responded incorrectly, and for each incorrect response, to move up one level of shock on the generator. In the original set of studies, an authority figure in the guise of an official looking "researcher" in a lab coat was positioned behind a desk in the same room as the teacher, and the learner was positioned out of sight in another room. The learner began to complain and then shout discomfort as the voltage increased and eventually became completely silent after the 300 volt mark. Obedience was measured by how far the teacher would increase the level of shock before refusing to continue. Nobody refused to continue prior to the 300 volt mark. Nobody. Sixty-five percent of the participants went the full distance to the 450 volt mark. Milgram did a variety of follow-up studies varying the conditions. The most relevant findings in these follow-up studies for our purposes here are (1) that the highest level of obedience occurred when the learner and teacher were isolated in different rooms and the learner could not be either seen or heard (93% went to the top of the voltage scale), and (2) when the learner and teacher were in the same room and the teacher was required manually to force the learner's hand onto a shock plate, the rate of obedience dropped to a mere 30%.

These latter two findings from Milgram's data raise some obvious questions with respect to the exploding level of technological mediation in our present society. How is our willingness to cause emotional injury to others increased by the distancing effects of email and text messaging? What does it say about the potential future applications of drone military vehicles and aircraft if obedience to commands to inflict pain is enhanced by isolation from the target of your aggression? In the past, a soldier had to physically drive the sword through his enemy's ribs while looking him in the eye and hearing him scream; now it is just a video game where soldiers of both genders maneuver joysticks and push buttons while monitoring a video screen—literally child's play. Because of advances in technology, commands from above are becoming easer to follow, psychologically as well as physically.

Another factor that limits resistance to authority is *perceived consensus.* Perceived consensus is a powerful source of personal belief and an explanation for the persistence of irrational beliefs in a society. Consensus, or the perception that there is a general consensus, "creates a sort of normative inertia: Consensually shared beliefs and behaviors tend to be self-replicating" (Conway & Schaller, 2005, p. 311). Not all forms of consensus are equal, however. It is possible for persons in positions of power and authority to compel consensus through direct or implied threat. Obedience-based consensus is extremely fragile, and wide-spread deviance is virtually guaranteed as soon as the threat is removed. Obedience-based consensus is particularly fragile when the authority's authority is based on power rather than on expertise (Conway & Schaller, 2005). The power of consensus to maintain belief and behavior is tied to the perception that the consensus truly represents popular opinion and not the enforced will of an authority. Thus an authority's commands can backfire, and research in this area suggests that it is possible to undermine the power of consensus by simply creating the perception that it is linked to the will of an authority figure (Conway & Schaller, 2005).

The field of social psychology abounds with evidence that dominance relationships can be elicited as a function of social roles which can be arbitrarily assigned (e.g., Zimbardo, 1973). Dominance in one situation does not preclude submission when the context changes, and vice versa. Research has found that when formerly oppressed groups occupy positions of power and authority, they tend to react to their new-found dominance in a way that suggests a rebound effect: they abuse their new power through displays of in-group favoritism and out-group hostility; and this tendency to abuse power is exacerbated when perceived control is low, and most pronounced when the abuse is socially endorsed (Prislin, Sawicki, & Williams, 2011). Abuse of power by a formerly oppressed group is more likely when there are social rewards for doing so and when the group believes that, although it is in a position of dominance, it has little ability to actually control things.

Authority and control are not synonymous. Occupying a position of social or political power gives one the ability to influence the actions of others, but having influence does not mean that one has control over the outcome. Psychologists have found, however, that those in positions of power are prone to develop an illusory sense of control in which they come to believe that they have personal control over chance events that are entirely outside their reach (Fast, Gruenfeld, Sivanathan, & Galinsky, 2009). Presumably, the greater the power, the further this illusion of control extends. The pharaoh may have actually seen himself as a god. The pope may come to believe he truly is a divine conduit. The corporate CEO may really think the structural integrity of the mine shaft or oil rig can be maintained by power of will, and so neglects to have proper backup systems put in place. This illusion of control is analogous to that of the small child whose interpretation of events is generated from within a restricted egocentric perspective. Along with the illusion of control, the obsession with power that is prevalent at the top of the corporate and political pyramid, the drive for wealth and fame, the fantasy for overwhelming power and dominance that serves as the bedrock of our consumer culture, is strong evidence for psychological unbalance, and reflects the infantilizing effects of arrested development.

Psychologically healthy people have no need to indulge fantasies of absolute power; nor do they need to come to terms with reality by inflicting self-mutilation and prematurely courting death. But the critical weakness of an over-regimented institutional structure—and almost by definition 'civilization' was over-regimented from the beginning—is that it does not tend to produce psychologically healthy people. The rigid division of labor and the segregation of castes produce unbalanced characters, while the mechanical routine normalizes—and rewards—those compulsive personalities who are afraid to cope with the embarrassing riches of life. (Mumford, 1966, p. 226)

Even in situations where people are forced to interact in overtly mechanical ways, such as occurs with workers on a factory assembly line or mid-level corporate bureaucrats, there are numerous facets of behavior that are external to the mechanical process, behavior that extends beyond the limits of the role, behavior that has the potential to reduce productivity and efficiency. People spontaneously engage in helping behavior, for example, sometimes even at the expense of their own job performance. And because this kind of behavior falls outside the formal role requirements it is not easily controlled or enforced (Smith, Organ, & Near, 1983). This is true in specific job situations as well as in the society at large. Research suggests that there are at least two categories of this type of "good citizen" helping behavior: one where the behavior is based on altruistic motives, the situation-specific assistance of people who are in need of help, and one where the behavior is based on a norm-based "general compliance," helping others for the good of the system (Smith, Organ, & Near, 1983). There is evidence that people with a high need for approval are more susceptible to general compliance norms (Smith, Organ, & Near, 1983). Note that the need for approval is highest in children and adolescence and tends to decrease with maturity (Bishop & Beckman, 1971). The implication here is that, paradoxically, the integrity of the system benefits from the psychological immaturity of its participants.

Whether someone is more likely to engage in helping behavior for altruistic motives or for general compliance reasons—or at all—depends to some extent on whether they are living in a city or in a more rural setting. City dwellers are less likely to offer assistance of any kind, and most likely to do so as function of general compliance. It has been suggested that urban inhabitants may suffer from a stimulus overload that decreases their sensitivity to the needs of others (e.g., Smith, Organ, & Near, 1983). The continual overstimulation of sensory and perceptual systems may lead to an overtaxing of perceptual systems and a subsequent reduction in neural sensitivity that prevents the perception of subtle signals, or perhaps the sensory demands of urban environments may simply out-compete human-generated signals for attention; recent neuroimaging studies that show that the brains of city dwellers respond differently to stressful stimuli than do the brains of their rural counterparts (Abbott, 2011) suggest that these two possibilities are not mutually exclusive. Cities are now the dominant form of human habitat: world-wide, the rural-urban balance shifted in just this century, specifically, on May 23, 2007, according to one study (Wimberley, Fulkerson, & Morris, 2007). Here in the US, according to census data, roughly 80% of Americans live in urban areas; and almost 60% live in cities with a population of 200,000 or higher. We might expect to see a continued erosion of altruistic behavior world-wide as the urban to rural ratio increases.

A person's identity is a function of their relationships with other people. Because we are forced to align our interpersonal interactions with the mechanical flow of power through the invisible machine of civilization, we are led to the formation of predominantly superficial rather than deep and meaningful relationships with other people. The frenetic pace of city life exacerbates this tendency, and our identities become thin and brittle.

Many aspects of the urban hive are shaped by the industries of transportation, energy use, and state-of-the-art synthesis of materials and products. On the other hand, the city is shaped, designed consciously and unconsciously, by identity cripples, deprived in various social and ecological dimensions, yet also cripples in the sense of potential capacity, the possibilities of personal realization in the archaic and magnificent environments of the deep past (Shepard, 1982, p. 121).

Modern life severely inhibits the ability to establish complex and meaningful relationships with others. A logical consequence of this is an increase in the prevalence of loneliness, a condition of deficiency that spreads very much like an infectious disease.

Loneliness does not have a bipolar opposite like happiness, but, rather, is like hunger thirst and pain [...] Furthermore, as an aversive state, loneliness may motivate people to seek social connection (whatever the response of others to such overtures), which has the effect of increasing the likelihood that those proximal to a lonely individual will be exposed to loneliness. Together, these processes may make loneliness more contagious than nonloneliness (Cacioppo, Fowler, & Christakis, 2009, p. 988).

Loneliness, which is a perception and not an objective condition, has a more profoundly negative impact on physical and psychological health than does actual social isolation (Cacioppo, Fowler, & Christakis, 2009). Also, research suggests that loneliness is a condition that spreads like a social contagion; loneliness spreads in social networks from person to person though direct social contact (Cacioppo, Fowler, & Christakis, 2009):

[W]hat might appear to be a quintessential individualistic experience—loneliness—is not only a function of the individual but is also a property of groups of people. People who are lonely tend to be linked to others who are lonely, an effect that is stronger for geographically proximal than distant friends yet extends up to three degrees of separation (friends' friends' friends) within the social network. The nature of friendship matters, as well, in that nearby mutual friends show stronger effects than nearby ordinary friends (p. 985).

Think about this in terms of recent innovations in social media. We've already discussed the fact that social isolation has been increasing as of late, and that it has as a potential source the very communication technologies that are advertised to enhance and expand our interpersonal relationships. What does the future hold for a technologically interconnected society if loneliness is contagious and spreads through social networks where we all have mutual virtual friends? Might our social media networks serve, paradoxically, as a global vector for a pandemic of loneliness?

Surely in time we will adapt to our changing social circumstances. Humans are among the most adaptive creatures on the planet, second, perhaps, to bacteria. Civilization itself is ample evidence of human adaptability. Modern urbanites are the paragon of adaptive organisms. Or, conversely, does civilization represent an abject failure of adaptation. How is it adaptive to adopt a lifestyle that ensures your eventual demise? How is it adaptive for a species to permanently destroy its future habitat? The inhabitants of modern civilization reflect only an ostensive "[…] adaptability, that is, in the sense of flexibility, readiness to change jobs, addresses, beliefs, celebrated by the technocratic ideal of progress in convenience, comfort, safety, insulation, and the stimulus of novelty. This kind of adaptability is not of a citizenship that transcends place and time, but of not yet being adapted, of never finding one's place or time" (Shepard, 1982, p. 121). What we think of as adaptability is only our programmed readiness to acquiesce to the shifting demands of the machine, our willingness to align our daily lives with the flow of power, our eagerness to abandon our humanity for the chance to ratchet up to a higher level of dominion, to occupy a more central role in the schematic layout of the megamachine.

A Thought Experiment

As a transition to the discussion of resistance that follows in the next section, I want to end this part with a thought experiment involving the comparison of two scenarios, one dark-dystopia science fiction, and one real.

Scenario 1: Imagine a technologically sophisticated civilization of about seven billion people. Imagine that this civilization has managed to solve most of the big problems that affect us in our own global technologically sophisticated civilization, including the problem of energy generation. Imagine that this civilization has discovered an unlimited source of environmentally-friendly renewable energy. Imagine that the power source is human beings, specifically a novel kind of energy released through the painful and slow incineration of human beings while they are still alive. Assume for thought-experiment's sake that no other animal will work, it has to be living humans. Imagine that it takes the combustion of several million humans each year to meet the massive energy demands of this civilization, but since infertility is low and most disease has been eliminated, there will always be plenty of people to feed the furnaces. The process is completely sustainable. Imagine further that this society has arrived at a completely equitable method of determining who ends up in the furnace, perhaps a lottery in which each and every living person has an equal chance of being selected on any given day. Now for the thought experiment part, what would it be like, from a psychological standpoint, to be a citizen in this world, knowing that any day you or your children might be hauled away and fed into the furnaces, knowing that in order for you to toast your bread in the morning someone has to die a slow and agonizing death? What kind of moral monsters would people have to become in order to accept this kind of an arrangement?

Scenario 2: Imagine our own technologically sophisticated civilization, one in which tens of millions of people are killed each year in order to support our industrial infrastructure. Technology kills people directly through car, train, boat, and plane wrecks (I refuse to call them "accidents" because, although they are not predictable, they are expected and unavoidable results of mechanical failure and operator error), industrial mishaps, bridge and building collapse, appliance malfunction or misuse, and through the deliberate use of military weaponry. The latter category is by far the largest, and is likely to grow larger in inverse proportion to the decrease in available fossil fuel, which is itself a result of energy-consuming technology. These deaths are not avoidable, or even *theoretically* avoidable. They are part and parcel of how our technological system works. Technology also kills people directly in less observable ways, such as through exposure to industrial carcinogens, through the chemical alteration of the air, water, and soil, and through metabolic and vascular diseases of all flavors caused by processed food and technology-mediated lifestyles. Add to that drug and alcohol use, homicide, suicide, and diseases that would never exist if they did not have densely populated urban settings in which to incubate, mutate, and spread.   Now back to the thought experiment part. What are the essential differences between scenario 1 and scenario 2? In both scenarios millions of people have to die each year in order to power the machine of civilization, in order to maintain the technological infrastructure. The most obvious difference is the lack of explicit acknowledgement (or even conscious awareness) of the relationship between technology and death in scenario 2. Another way that the second scenario differs from the first is that in our society the selection process is not random. The method of choosing who dies on a given day is not equitable and is heavily weighted toward certain groups of people: you are more likely to die as a direct result of technology if you live in the third world, if you live below the poverty line, if you are a person of color, or if you are a child.   What would it be like to live in our own civilization as a fully aware and conscious human being? What would it be like knowing that in order for you to toast your bread in the morning someone has to die a (perhaps) slow and agonizing death? What kind of moral monsters would we have to become in order to accept this kind of arrangement?

As an addendum to our thought experiment, compare scenario 2 with a scenario based on the discussion of the nature of "primitive" foraging culture presented in Part 1.

> From the standpoint of human survival, to say nothing of further development, a flint arrowhead is preferable to a hydrogen bomb. Doubtless it hurts the pride of modern man to realize that earlier cultures, with simpler technical facilities, may have been superior to his own in terms of human values, and that genuine progress involves continuity and conservation, above all, conscious anticipation and rational selection—the antithesis of our present kaleidoscopic multiplication of random novelties. (Mumford, 1974, pp 203-204)

PART 3: RESISTANCE

The Mantra of Power: Resistance is Futile

Fetters

Rousseau surely overstated when he declared that "Man is born free, but everywhere he is in chains." We are tethered to the machine even in the womb, where the winnowing of future possibility begins with hormone-disrupting pesticides, heavy metals, an alphabet soup of industrial toxins, and the chemical shock of neonatal vitamins. And in the first hours and months and years, the critical dance of maternal attachment, the back-and-forth, give-and-take of early identity formation becomes a stuttering, frame-jumping mélange, shrouded in sensory deprivation and interleaved with commercial advertising. As we mature, complete maturation never happens, and our cyclically-expanding epigenetic emergence becomes a linear trajectory bent to the ends of an invisible mechanical leviathan that is even now chewing the ground from under our feet.

Successful resistance requires at least four things. First and foremost is the recognition that there is something to resist against; knowledge of bondage is an obvious precondition. A slave who believes that her servitude is a noble calling has no thoughts of resistance. Second, knowing the means by which freedom is restricted, understanding the nature of our restraints is vital for choosing potentially effective courses of action. As with quicksand, intuitive reactions in the absence of a clear understanding of what we are dealing with can make our situation worse very quickly. Third, selecting the appropriate target of resistance is also essential. The stage magician's kit of redirection, diversion, and distraction, is used effectively by military strategists, corporate advertisers, and political propagandists as well, and can easily confuse our aim. And while we expend our energies on effigies and red herrings, our chains are fortified. An additional difficulty is that we are facing a many-headed hydra, an extremely complex and multifaceted oppressor armed with ever-changing continually-upgraded self-preservation technology. And finally, it is necessary to choose courses of action that have some actual likelihood of yielding positive results. Pleas for compassion, for instance, have little chance of success if your oppressor is an unfeeling machine.

We have already discussed several ways in which our bondage to the machine of civilization is obscured from us. The very immensity and complexity of our global techno-culture prohibits any meaningfully coherent perspective. Specialization and social stratification, the careful partitioning of knowledge and power, the isolation and insulation of information prevents an integrated comprehension of even minor facets of the machine's operation. And we learn from childhood not to question the legitimacy of the status quo, to view civilization as benevolent, and to view its negative features as unintended results of human imperfection or as temporary situations that will be dispelled by future progress. Belief in progress, the myth of linear progression from the torpor and sloth of primitive ignorance to the active technological engagement of civilized enlightenment, takes on the inertia of delusion, unaffected by mountains of evidence to the contrary. Civilization is inevitable; it is our species' manifest destiny. We come to believe that our best interests and the best interests of civilization are the same, and align our personal goals and aspirations accordingly. We come to identify with our captor and see civilization as an extension of ourselves. I remember watching the first moon landing on television as a young child; I shared in the pride that I felt emanating from the adults in my life at the time. *We* did this. *We* conquered the emptiness of space and pinned a flag on the moon. It was only several decades later that I began to question just how non-voluntary my affiliation with this *we* actually was. But my questioning was—and still is—an aberration, a statistical anomaly. The vast majority of the machine's human victims never have any reason to question. Civilization is simply who we are.

Even after one begins to question the ultimate benevolence of civilization, there are several psychological hurdles that must be cleared before any serious thought is given to possible alternatives. The *existence bias,* the tendency to base judgments of goodness and value on the bare fact that a situation already exists (or on the belief that it exists) is one such hurdle.

> People infer goodness from existence. [...] In consequence, alternatives to the status quo are disadvantaged; more effort will be required to evaluate them positively. Because of a bias favoring mere existence, not only is it difficult to effect social change but it can be difficult simply to persuade people to think favorably about alternatives. (Eidelman, Crandall, & Pattershall, 2009)

Civilization isn't perfect, but it's what we've got, and that fact alone makes it preferable to potential alternatives. The psychological deck is stacked in the favor of preserving the status quo.

The existence bias is an example of a heuristic. Heuristics are mental short-cuts that generate appropriate conclusions under normal conditions, but can on occasion lead us wildly astray (Tversky & Kahneman, 1974). The cognitive load imposed by the demands of civilized life, through multi-tasking, anxiety and stress, and persistent distraction, increases our reliance on heuristics and makes it less likely we will take time to think rationally about our situation, less likely that we will notice—let alone take umbrage to—the involuntary nature of our daily activities.

Along with the existence bias, and a variety of additional built-in cognitive blinders that lead us toward a passive acceptance of the status quo, multiple layers of accommodation to technology have generated a deep dependency on the machine of civilization, a dependency that shares a surprising number of features in common with the physical and psychological dependence that attend drug addiction. Viewing our situation as a kind of addiction provides potentially useful insight, not only for understanding the psychological nature of our chains, but for mapping out potential "courses of treatment" as well. Rationalization, denial, avoidance, and other defense mechanisms commonly employed by addicts to justify their continued drug use also play a role in our psychological bondage to civilization. Drug addiction is a treatable condition, however, and there is an extensive arsenal of effective treatment methods and an extensive literature on the recovery process. I will explore how our understanding of drug dependence might be applied to the treatment of "civilization addiction" in some detail below. From a psychological perspective, civilization dependence is, theoretically at least, treatable.

Our principle and most durable chains are those of arrested development. Psychological modules designed to assemble in response to social and material conditions distinctly different than those we actually experience assemble themselves nonetheless. But they are unfinished and distorted. And their integration with other modules is incomplete, resulting in a permanent and systemic deficiency requiring external fortification. Because of the mismatch between the input expected by our epigenetic programs and the input provided by civilization, civilized adults are developmentally incomplete, emotional infants in a perpetual state of psychological need. Civilization exploits that need in innumerable ways, channeling desire and action in directions that service and expand the movement of the machine. Our immaturity is the source of a baseline insecurity that provides the machine with sturdy psychological carabiners to latch onto. And the carabiners quickly become shackles. Civilization both creates and services our insecurities. Civilization first becomes an oxygen tent, and then an iron lung. Arrested development exacerbates the effects of a variety of psychological snags that keep thoughts of resistance from getting any solid purchase.

Arrested development engenders child-like, ego-centric perspectives in which we see the world narrowly in terms of ourselves. We have already looked at the tendency to reify, to treat abstractions as if they had an independent concrete reality. This tendency is enhanced by the infantile projection of agency onto inanimate features of the world.[4] As a result, institutions become personified. We talk of governments as if they were goal-directed beings with personalities, as if they were deserving of our worship or derision. Corporations, abstract legal-economic systems, are granted person status and compete with biological people in a grossly lopsided contest for access to resources. Corporations have replaced god-kings at top of the modern megamachine's power hierarchy. And, like the pharaohs of old, the legitimacy of their power is unquestionable, a matter of other-worldly decree. The personification of corporations frames our understanding in ways that limit our ability to see the actual role these powerful institutions play in keeping us passive and compliant. But personification, the metaphoric application of human attributes to nonhuman entities, can work for or against the status quo: personification is a potential linchpin in the system, providing potential traction for resistance. What happens when we start to treat corporations as if they were actual people—people who have become tyrants demanding our complete subservience, people who have enslaved us, robbed us of our birthright as human beings? What would it mean for the masses to rise up and issue a death decree for civilization's most powerful institutions? I imagine Bastille-day style executions in which the material substance of the global corporate world falls under the massive guillotine of wide-spread popular resistance.

It should be clear at this point that the target of our resistance is not human, but a hyper-complex machine that is systematically exploiting our evolutionary-derived dispositions for a completely different kind of lifestyle than the one civilization has to offer. Unlike a relatively simple physical machine, say, an automobile, the machine of civilization is an exceedingly vast and diffusely organized collection of interdependent systems that include uncountable physical and conceptual objects, an enormous variety of organizational structures, and the complex networks of interaction that unite them. Our fetters are manifold and multifarious; there is no obvious single focal point of attack. An automobile or a clock can be rendered nonfunctional by simply removing, deforming, or destroying a critical component part. The vulnerabilities of civilization are concealed within the dense complexity of the system.

There are some who have suggested that the system's complexity is itself a potent vulnerability. For example, Mackenzie (2008) suggests the complexity of our global civilization makes it vulnerable to relatively minor work-force disruptions, and that a work-force reduction similar to that caused by the flu pandemic of 1918 could terminally derail the entire global system. The weakness is a matter of interdependency and a complete lack of redundancy. The high degree of specialization in our globally-networked civilization leaves the entire network at the mercy of its weakest nodes. There are two interesting things about this notion. First, the division of labor and specialization that serve as fundamental principles of civilization are also a lethal vulnerability, perhaps the very source of its eventual collapse. Second, it suggests that civilization is becoming increasingly fragile, which offers some hope to those who choose to resist; and, as with simpler machines such as an automobile or a watch, there are numerous potential specific targets for resistance. Selecting the right target might not be as important as simply making a commitment to resist.

So what keeps us from resisting the machine?

Technology as a Monkey Trap

Our relationship with the corporate world shares much in common with Stockholm syndrome, the name given to the supposed condition in which hostages come first to sympathize with, and then to identify with, their captors. The men and women sitting in corporate board rooms control, directly or indirectly, virtually every non trivial aspect of our lives. And they do so through psychological manipulation, coercion, and the direct threat of violence. We are hostages. We have no choice but to participate in the corporate game. Even those of us who have some understanding of the dominating role that powerful corporations play in our lives continue to support them directly by purchasing their products and services and indirectly by our passivity, by allowing them to exploit the commons and by failing to demand restraint when they engage in practices that threaten the health of people and the natural environment. Rather than wallow in our own feelings of helplessness, we come to sympathize with our oppressive captors. We adopt their ends as our own. We go so far as to proudly wear corporate logos emblazoned on our clothing—and willingly pay for the privilege of serving as walking advertisements.

In terms of personal freedom, corporate privatization and the ideology it promotes leads to a paradox that Benjamin Barber (2007) refers to as a "civic schizophrenia," in which an individual's interests as a private person are placed in conflict with her interests as a public citizen:

> Privatization ideology treats choice as fundamentally private, a matter not of determining some deliberate "we should" (a kind of "general will" produced by citizens interacting democratically) but only of enumerating and aggregating all the "I want's" we hold as private consumers and creatures of personal desire. Yet private choices do inevitably have social consequences and public outcomes. When these derive from purely personal preferences, the results are often socially irrational and unintended: at wide variance with the kind of society we might choose through collective deliberation and democratic decision- making. (p. 128)

Privately, I want my big screen high definition television, my high-speed internet, my smart phone, my weed-free lawn, and the freedom to drive my SUV to the local box store where I have access to cheap imported consumer goods. As a public citizen, however, I want to live on a planet with clean air and water and in a neighborhood with character, a low crime rate, and a vibrant local economy. Thus our corporate consumer system sets up a kind of social trap in which our private interests are pitted against our public interests, a trap in which we are coerced to participate in dehumanizing and environmentally destructive activities.

John Platt defines social traps as:

> situations in a society that contain traps formally like a fish trap, where men [sic] or organizations or whole societies get themselves started in some direction or some set of relationships that later prove to be unpleasant or lethal and that they see no easy way to back out of or to avoid. (1973, p. 641)

Probably the best known description of this kind of trap comes from Garret Hardin (1968) and his discussion of the tragedy of the commons, a tragedy that results when individual interests and collective interests are mutually exclusive. Briefly, the tragedy of the commons refers to a situation in which villagers can freely graze their livestock on a common pasture, and it is to each individual villager's advantage to graze additional livestock even though doing so will lead to the overgrazing and eventual destruction of the pasture. It is a situation where community interest and individual self-interest collide. Maybe it is better thought of as a conflict of self-interests, a situation in which short-term self-interest (more livestock for me and my family to live on) conflicts with long-term self-interest (future pasture on which to graze the livestock). The tragedy is that if I don't use it, someone else will, and then I sacrifice both short- and long-term gain, so the only rational action for me to take is to put as many animals on the commons as I can now. Unfortunately, that's also the only rational action for my neighbors to take as well, so the pasture is quickly overpopulated. The tragedy of the commons can be easily applied to numerous specific situations involving the extraction of natural resources or the polluting of air, land, and water by both individuals and industrial corporations. The magnitude of the force and scope of this kind of trap is increased exponentially when combined with a mass-technology consumer-based global civilization such as ours.

This idea of a social trap goes a long way toward explaining our increasing dependency with respect to technology. Automobile transportation serves as one of several potentially salient examples. Initially, the automobile was an innovative alternative to horse-carriage transportation. The widespread adoption of cars led to the construction of a complex highway system, the dispersal of the population, and the practice of commuting distances that would not be practical otherwise. In addition, a massive industrial infrastructure dedicated to the manufacture and fueling of these vehicles has emerged and become so deeply integrated within all parts of the system that putting an abrupt end to the use of automotive transportation is unthinkable, even when it is obvious that the entire biosphere is buckling under the weight of the environmental havoc it causes. Our entire social fabric has been woven around the automobile to the point where any threat to the continued availability of this technology risks unraveling the entire system. We willingly destroy sensitive ecosystems and engage in war to maintain our supply of oil. We are trapped. A very similar scenario could be constructed for our increasing dependence on of any of a number of more recent technological innovations, computer technology or cell phones, for example. Technology becomes a kind of monkey trap in which we are unable to let go even when the problems become enormous. In fact, the more extensive the problems, the more highly interpenetrating the technology, the more tightly we hold on.

In my discussion of the invisible machine I have been careful not to equate civilization with technology itself. The megamachine of civilization is not some systematic aggregate of technology; nor is it some sentient agency using technology to control our lives. The megamachine employs technology, but it is itself a kind of technology. The conditions that bring about and support civilization—division of labor, specialization, and the uneven distribution of power—are also the requisite conditions for a technologically-embedded lifestyle. Specialization and the isolation of knowledge are especially important. Technological innovation is highly domain-specific, typically focused on a single set of problems in a highly circumscribed area. It is this isolation of knowledge and activity that prevents even the most peripheral attention to distributed consequences of technological innovation, consequences in areas not directly related to the set of problems the innovation was designed to address. Distributed consequences are the potential focus of specific future innovations, which then yield additional consequences. And the problems of technology continue to snowball in this fashion, not because technology has somehow gotten away from us, not because it has a mind of its own, but because the technological process is, by its very nature, one that is entirely unresponsive to feedback from distributed or long-term outcomes. Langdon Winner (1977) refers to this set of circumstances as "technological drift." Lifestyle changes generated to accommodate the application of specific technologies (think of automobiles here) create new problems and set the stage for the development of future innovations that necessitate further lifestyle changes. Several generations down the road, we find ourselves deeply dependent on a complex web of technologies and committed to ways of life that we would have never chosen if we were given the choice at the outset. Technology is quicksand that that keeps us eternally thrashing with short-term fixes that drive us deeper and deeper into the mud.

This suggests that it is unlikely that there is a technological solution to the problems of technology—and by extension, to the problem of civilization itself. Technology is a monkey trap, and grasping at technological solutions only keeps our fist pinned inside the jar. We need to learn the art of letting-go. But although technology is the proximal source of our pain, it is not the enemy. Successful resistance is unlikely to consist (entirely) of direct attacks on specific technologies. To be effective, resistance needs to have an impact on the underlying source of the problem: the enabling conditions: the division of labor and the hierarchical power relationships and isolation of knowledge it engenders. Neither specific technologies nor civilization as a whole can function in the absence of these enablers.

Overwhelming Force

There are two rats. The first rat is put in a cage in which the floor has been wired to produce a painful electric shock. In the cage is a small wheel that, when turned, will break the circuit and stop the flow of electricity to the floor. An alarm tone sounds, and after a short delay the floor is electrified. It takes only a few trials before the rat learns that it can completely avoid the shock by turning the wheel as soon as it hears the tone. The second rat is put into the same cage, only this time the wheel is not connected to the circuit, and spinning it has no effect. There is no way to avoid the shock, and the rat learns to freeze and brace for the inevitable when it hears the tone. Later on, both rats are introduced to a different cage, this one divided in the middle by a wall with a small rat-sized opening. The floor on each side of the cage can be electrified separately. The first rat is put into the cage, the alarm is sounded, and the floor on that side of the cage is electrified. After just a few trials, the rat learns to shuttle into the other side of the cage as soon as the alarm sounds—and learns this faster than a third rat who was not involved in the first part of the experiment. What does the second rat do when placed in this new cage? It freezes at the sound of the alarm and braces for the inevitable shock. Even after numerous trials, rat number two never learns how to avoid the shock.

This behavioral experiment reflects a phenomenon called *learned helplessness.* Learned helplessness occurs when an organism learns that there is a lack of contingency between responding and an aversive stimulus: the aversive stimulus will occur no matter what, and the organism is helpless to avoid it. The critical issue with learned helplessness, highlighted in the shuttling experiment, is that it can carry over to situations other than the one in which the original lack-of-contingency learning occurred, and prevent the organism from taking action in situations where there is a contingency, where there is something it can do to avoid or escape the aversive stimulus.

Learned helplessness has been generated in humans under a wide variety of laboratory conditions (Seligman, 1975). Once established, learned helplessness is difficult to overcome. But it is possible to become immunized against learned helplessness by experiencing success with a controllable task before facing the one which is uncontrollable. Note that the first rat in the example above learned to shuttle to the non-electrified side of the cage more quickly than did a naïve rat who did not participate in the wheel-turning part of the experiment. Prior success at discovering one set of contingencies appears to enhance the ability to discover a second set of contingencies in different but related circumstances. Even more important, research in this area has found that previously controllable experiences appear to have a prophylactic effect and interfere with the establishment of learned helplessness to begin with. Further, this immunization effect is enhanced in humans by attributing failure in uncontrollable situations to specific external sources (Ramírez, Maldonado, & Martos, 1992; see also Mikulincer, 1986). That is, you are less likely to see yourself as helpless in a situation if you have had prior success in a similar situation and if you attribute failure to a specific source outside of yourself (e.g., "The task was purposely designed to be too difficult to accomplish," or "Some obstacle prevented me from succeeding"). Just because you may have been successful at overcoming a challenge in the past, however, is no guarantee against acquiring learned helplessness in a future, unrelated, situation. Simply thinking about past successes prior to confronting an uncontrollable situation does not appear to have a strong immunization effect against learned helplessness (Teasdale, 1978). There are numerous real-world situations that promote learned helplessness, no-win situations in which you are "damned of you do, damned if you don't." Learned helplessness is often present in abusive relationships and can occur in poorly structured educational settings, and it has been linked to depression (Seligman, 1975), and to a variety of cognitive and emotional deficits (Maier & Seligman, 1976). Learned helplessness is perhaps the most significant psychological threat to meaningful resistance.

Resistance is essential in a machine made of metal. Without contact between resistant surfaces, without pushback and friction, movement of the gears would be impossible. But resistance has to be tightly controlled and kept within a specified range of tolerances. The machine of civilization is likewise dependent on the presence of a certain range of controlled resistance among its component parts. But unchecked resistance is not "tolerated." Open refusal to comply with the machine's operation, either by resisting in ways that threaten the structural integrity of one of the machine's components or by simply refusing to appropriately engage the system's powertrain, is a potential death sentence. Note that "refusing to comply with the machine's operation," is not the same as "breaking the law." Crime is part of the machine's design, and criminals serve a variety of vital functions. A thief, for example, by his or her very act of theft, demonstrates open deference to the idea of private property. The risk of theft also provides justification for police "protection," surveillance, and other restrictions on personal movement and privacy that enhance the machine's control. Also, criminals make excellent diversionary scapegoats, redirecting attention and preventing scrutiny of the system itself: society's problems are not part of the nature of the system; rather they reflect the deviant activity of a criminal class of people. Drug dealers, arsonists, vandals, shoplifters, child molesters, fraudulent realtors, and axe murderers are as much part of the machine as are lawyers, ballet teachers, and air traffic controllers. But by stepping outside of the machine, by refusing to acknowledge the machine's legitimacy, by resisting engagement in the machine's operation, you become like the rest of the natural world: external material to be either exploited and consumed or eliminated. If you refuse participation, if you offer nothing of value, then, from the machine's perspective, elimination is the only option.

It rarely comes to that, however, because true resistance seldom occurs. And when it does, the media machine is careful to recast it in terms of criminal psychopathology or screen it entirely from public awareness. The modern megamachine employs several effective mechanisms for ensuring unquestioning acquiescence to its operational demands. Some of these mechanisms are overt and transparent, such as the designation of "authorities" tasked to maintain, enforce, and express the machine's monopoly on violence: the whole of the criminal legal system, the police, covert government organizations, and the military. Some mechanisms are more subtle, such as the propaganda generated by the media and the entertainment industry, both of which operate under the same prime directive: homogenize values and manufacture consensus. The former mechanisms, the use and threat of violence, were more visible, perhaps, in earlier manifestations of the megamachine, when the components of the labor force were entirely organic and overt slavery was a dominant mode. In the modern iteration, violence has become deemphasized as a control mechanism—it's a kinder, gentler machine. But violence still forms the core skeleton of the modern machine, every bit as much and even more than it did the earlier versions; the accumulation of technological innovation has also been an accumulation of new and more efficient methods of violence.

Violence forms the core of the modern megamachine, but the machine's invisible exoskeleton is composed of carefully integrated deception and psychological sleight of hand. In addition to ubiquitous propaganda, the machine keeps us in check by convincing us that we are powerless to resist. It accomplishes this by subverting and conditioning our psychological predispositions in at least three interrelated ways that encourage learned helplessness: by capitalizing on the dependency that stems from arrested development, by limiting our ability to self-regulate, and by maintaining a complete monopoly on violence.

Arrested development leaves us perpetually immature and vulnerable to external control. Resistance to authority, when it does occur, resembles the petulant power-struggle of an adolescent trying to wrangle the car keys for a Saturday night out rather than a struggle for true independence. The deep dependence that goes along with our arrested maturation also exacerbates our sense of helplessness. Human infants are in fact helpless. They need adult care and protection in order to survive. Likewise, our dependency, our lack of self-reliance, leaves us as infants in the care of the machine. We are taught that we need civilization and believe that we would quickly die without its continual nurturance. The idea of living entirely outside of civilization's embrace is as unthinkable as it is rapidly becoming physically impossible. Also, the narrow self-focus and impulsivity associated with infantile modes of being keep the vast majority of us far too engaged in the superficial distractions offered by the machine to consider resisting its machinations; we are too firmly attached to the machine's teats.

Along with encouraging and rewarding infantile impulsivity, a second and related way that the machine keeps us helpless and passive is by restricting our ability to exercise self-control. A person with self-control is a potential threat to the machine. The extent to which a person is able to regulate their own thought, behavior, and emotional states is the precise extent to which they are free from external regulation. But we learn to fear people who think for themselves and act in ways that deviate from established norms. And a person who thinks and acts entirely for themselves is a monster, a wild, feral, uncivilized beast, in the eyes of the mindless, perpetually pacified, domesticated masses. Thus our behavior is firmly tethered to externally-imposed contingencies, and our physical and social environments are groomed in ways that limit our ability to self-regulate and exercise self-control.

Research on self-regulation has found that self-control operates very much like a muscle: it appears to be a limited resource that can be depleted and requires rest in order to reestablish its strength (Muraven & Baumeister, 2000). Self-control reflects the ability to inhibit behavior or override competing urges. The exercise of inhibition is effortful, and inhibiting one set of urges leaves less energy left over for another contemporaneous or subsequent set. People fail to exercise self-control following recent situations in which their self-control has been taxed or when there are multiple demands on their self-regulation that leave them with chronically deficient inhibitory resources (Muraven & Baumeister, 2000). For example, a dieter who has to inhibit the desire for the delicious chocolate chip cookies in the lunchroom is more likely to binge on cheesecake at home later on. Or the menial employee who has to submit quietly to a verbally abusive boss all day long on the job is more likely to lash out uncontrollably at his or her teenage child at the dinner table in the evening.

Muraven and Baumeister (2000) list several key features of self-control: (1) there is a common resource that supports widely different forms of self-control, (2) this resource is used by the part of the self that controls decision-making and behavioral initiation and control, (3) this resource is limited and can be completely exhausted, and a person can override only so many competing urges at the same time, (4) the decrease in self-control resource is not permanent, and can be replenished with rest in much the same way that strength eventually returns to a fatigued muscle, (5) self-control strength is variable among persons and situations, and, perhaps most notably, (6) it is possible to increase a person's self-control resource reservoir by frequent exercise of self-control followed by periods of sufficient recovery.

A wide variety of factors, environmental and psychological, can tap into this resource and reduce the ability to maintain self-control, including noise, coping with stress, regulating mood and moderating emotional expression, delaying gratification, dieting, and engaging in activities that require physical or attentional stamina; and there is evidence that just the anticipation of a future demand on self-control can deplete the inhibitory resource (Muraven & Baumeister, 2000). It is a fairly straightforward matter to trace the myriad ways that our self-control resource is overtaxed by life in modern techno-culture, with its continual stress, competing demands on attention, enticements, distractions, and a plethora of situations in which our natural emotional responses must be held in check. True resistance, then, requires an almost super-human level of self-control simply to overcome distraction and competing demands long enough to realize the extent to which resistance is warranted in the first place, let alone to meet the self-regulation demands of planning and executing effective courses of action and coping with their consequences.

There is power in numbers, and the physical numbers will

always be on the side of the status quo—and in an overwhelming proportion. For most people, the idea of resistance is unthinkable: to resist civilization is literally to unplug from life-support. Those few in whom the seeds of resistance have fallen on receptive cognitive soil thus have the additional burden of navigating the abrasive social and psychological environment of marginalization. Numbers behave somewhat differently in an individual's psychological calculus than they do in the concrete physical world, however. Over half a century ago, psychologists (e.g., Solomon Asch, 1956) showed that having just one ally was enough to break the all-too powerful spell of group conformity. So, for the seeds of resistance to take root in us as individuals, it is enough to know that we are not alone.

By the way: we are not alone!

But numbers do have consequences in the real world. Those who manage to clear psychological hurdles and social obstacles sufficiently to resist the machine will always represent a tiny minority. How can so small of a force have any impact whatsoever against such an enormous enemy? Even if real resistance is possible, it is unlikely to have any long-term success against such overwhelming odds. A third way the machine keeps us passive, one that makes maximum use of learned helplessness, is by exercising a complete monopoly on violence and establishing a policy of "overwhelming force" with respect to meaningful acts of rebellion. The policy of overwhelming force is the idea that all rebellion will be quashed regardless of what it takes to do so. The entire resources of the US National Guard will be employed to deal with a single resistant individual if need be. This is not a new policy, but has been part and parcel of civilization since the very beginning, because:

> Once a megamachine has been brought into existence, any criticism of its program, any departure from its principles, any detachment from its routines, any modifications of its structure through demands from below constitute a threat to the whole system. (Mumford, 1970, p.241)

The machine's directives will be followed no matter what it takes. Minor forms of protest are allowed as part of the need to maintain a minimum level of resistance in the system, but protesting has to be kept within strict limits; when it approaches these limits (which are a moving target that changes with cause and political climate), it will be met with violent and potentially deadly force. If a potential threat to the system is detected, it must be removed regardless of the cost or consequences in terms of "collateral damage." An animal right's activist who fire-bombs a research facility will receive a prison sentence far in excess of that warranted by the actual damage or threat to human life. Merely thinking about attacking the system, blowing up a railroad supply line to a coal power plant, for example, can lead to a lengthy prison sentence—or a one-way trip to Guantanamo Bay—if you happen to talk to the wrong person about it or describe your idea using the wrong words.

The policy of overwhelming force establishes the preconditions for learned helplessness. No matter how atrocious the machine becomes, no matter how heinous its actions are with respect to the environment or with respect to our humanity, we are helpless to confront the system directly. Our only option if we want to try to change things is to go through "the proper channels," which, of course, are channels specifically designed to ensure the machine continues with the minimal necessary friction. There is no way to redirect the machine from below by working within the system.

> Though even now few people seem to suspect the ideal form and ultimate destination of the industrial organization that has been taking shape in our own time, it is in fact heading toward a static finality, in which change of the system itself will be so impermissible that it will take place only through total disintegration and destruction. (Mumford, 1970, p. 211)

From a 21st century perspective, Mumford's words ring eerily prophetic.

The machine does violence to our human nature at its psychological source, and stacks the deck against resistance. But humanity has a couple aces up its sleeves. On a personal level, the potential for immunization against learned helplessness represents an exploitable weakness. So does the malleability of self-control. And there are numerous other targetable linchpins that emerge from the machine's failure to accommodate the mismatch between our ancestral heritage and current conditions. Although we are forced to assume the role of domesticated, mechanized cattle, we are still, in every cell of our cerebral cortex, authentic human beings, beings designed for a truly meaningful life embedded in the natural world. According to Paul Shepard (1998), it is important to distinguish between *domesticated* and *tamed*. Domestication involves the intentional alteration of the distribution of a species' genes to yield phenotypes that humans find desirable. In many cases, the result is an animal that is incapable of surviving without human care and intervention. Taming involves reducing the expression of certain subsets of instinctual behavior so that the animal can function in a context that is substantially different from its natural habitat. Humans—at least the currently living members of the species—are not domesticated in the biological sense. We still express many of the same phenotypes as those expressed by our foraging ancestors. But we are no longer allowed to develop our behavioral tendencies in ways that correspond to our genetic design. We have been tamed. What that means is that we are still wild creatures at heart. If there is any potential for successful resistance, it is tied directly to this fact; it is this fact more than any other that gives us hope for a more human future.

Rediscovering Human Nature

Rewriting the Myths of Civilization

The modern urban legend is a tale of progress, of inertial inevitability, of the beneficence of technological innovation, and of modern industrial civilization as a shining jewel at the pinnacle of our evolutionary trajectory. It is also patently false, a delusion. Unfortunately, delusions are notoriously difficult to dispel. It is not a simple matter of appealing to logic or presenting competing evidence. By definition, a delusion is a state of belief that is resistant to the intrusions of reality. Those who operate under the influence of a delusion are adept at dismissing inconsistencies among their beliefs and reality, and easily integrate competing evidence through a variety of very effective defense mechanisms.

The real problem with the myth of modern civilization is not that it is false, nor that it is unresponsive to a mountain of countervailing facts. The real problem is that the conceptual frame it provides leads us to act on the world in ways that are counter to our best interests—and to the best interests of every other life-form on the planet. A patently false system of belief that served as the grounding of a truly human way of life would not be a problem, despite its lack of veracity. The modern urban legend is a problem, but not because it is false. The modern myth provides validation for the modern version of the megamachine in much the same way that belief in divine kingship provided a source of justification for the destructive and oppressive activities of the first megamachines in Sumer and ancient Egypt several millennia ago. Note, however, that the causal arrow between the emergence of civilization and the development of a mythic rationale for its existence points in two directions. And it is only after civilization is firmly established that the myth becomes fully fleshed out. Thus the problem of modern civilization is not going to be solved by simply weaving another kind of story. Nevertheless, it might be useful to consider how we might reframe our situation in more human ways. It has been suggested that breaking a bad habit is not so much a matter of eliminating unwanted behavior as it is establishing new habits to take the place of the old. That is, an old habit never really dies; it is pushed into the background by the superposition of a new habit. Perhaps it is possible to encourage the formation of a new urban legend to replace the old. What kind of story might we compose about the conquest of the megamachine, what kind of mythic legend might we tell about our eventual rejection of civilization in favor of more authentic modes of living?

Perhaps we can look to the original legends of civilization, the Epic of Gilgamesh, or the Homeric tales, as a template. The legend of Odysseus, if I can be granted a wee bit of allegorical license, seems to provide an interesting pattern for a new kind of epic legend: a story of how the human species, after uncountable years abroad, eventually regains its place in the world. Odysseus' leaving home and going to war represents our species' transition from its evolutionary basis in foraging band society to lifestyles based on domestication and conquest, and all of the misery and strife that transition caused. Now, lost and under the beguiling spell of modern civilization, we find ourselves shipwrecked on an island with a powerful nymph, where life is a superficial paradise and yet we suffer chronic discontentment. For Odysseus, Calliope's island was a very empty place, and, of course, he was being held against his will. Despite its virtually unlimited array of momentary pleasures, modern techno-culture is hollow and unfulfilling; the island of civilization is not our home, and we are, like Odysseus, held hostage and forced to attend to desires that are not truly our own. And, like Odysseus' voyage home, our journey back to a life that is consistent with our human nature will be fraught with obstacles; there are storms on the horizon. And even if we manage to find our home shore again, ultimate success is not guaranteed. Things have changed since we left all those years ago. Our fortune has been squandered and our castle is crowded with men of bad intent. But we are not alone. We have our tirelessly devoted wife—the all-providing natural world which has never abandoned us. And we have our son—our genetic connection to the past—who has been abused, but is strong and itching for a fight. And, despite overwhelming odds, we yet possess the strength to string the bow and the steadiness to send the arrow on its narrow path.

But first we need to free ourselves from our seductress. We need to leave the island. Odysseus was able to break free only by enlisting the sympathy of his patron goddess, Athena. Athena no longer has ears for our kind. She is a goddess of war, and as such her sympathies are with the power complex of the machine. Who, or what, is our species' patron goddess? Throughout this book, I have been making the case that the science of psychology, named for the goddess Psyche, provides us with, if not freedom itself, some insights into the petitions that are necessary for us to begin to realize our own freedom.

And who, exactly, is our seductress? What is the source of her power over us? What is the force that keeps us bound to the island of civilization in the first place? We have explored some of the links in our psychological shackles, but what is their actual composition? What kind of blade is required to match their temper? In the next section, I suggest that potential insight into the answers to these questions can be gleaned from an unexpected source: research on the nature of drug addiction and recovery.

The Monkey on our Back

> People are content in the knowledge that things are as they are and that they are in good working condition, and that in society everyone and everything has a certain job to do and does it. And everyone, like every thing, does not find occasion to inquire into this condition or to dispute the manner in which it structures life. (Winner, 1977, p. 200)

People are willing to invest more energy and expense in maintaining the status quo, whatever the status quo happens to be, than they would have been willing to invest to bring those conditions about in the first place. Psychologists call this lopsided valuation of existing circumstances *status quo bias*.

In its simplest form, status quo bias, a relative of the existence bias mentioned previously, reflects a natural distaste for change, as if our present situation carries a kind of psychological inertia. It's easier to keep doing the same thing than it is to try something different—even if what we are doing isn't working out so well for us. So we put up with a job that is not entirely satisfying, or a marriage that is not entirely fulfilling. Our willingness to allow the continuation of our present civilization, however, is something more than just an aversion to change. It is something more closely akin to an addiction. It's as if we are addicted to civilization, hooked on consumption, and, as a result, entirely willing to ruin our personal health and the health of the planet in single-minded pursuit of our drug of choice.

Suppose that we could start from scratch, that we could reverse the clock back to the industrial revolution, or even as far back as the agricultural revolution, and plot the future of humankind from that point forward with full knowledge of what 21st century oil-dependent planet-devouring corporate consumer society would be like, with all of its violent, environmentally toxic, and inhumane accoutrements; that is, suppose that we had the option to *choose* our current civilization. Who would willingly make such a choice? It would be like asking a hopelessly addicted heroin junkie what she would do if she could return to the day of her first hit with full knowledge of where it would eventually lead, how it would come to control her life, destroy her most precious relationships, and ruin her health. Of course she would set the needle down and walk away. But now, as an addict, she is willing to go to extreme lengths, even to the point of selling her body, just to get her next fix.

In his State of the Union Address in February of 2006, then President George W. Bush invoked the metaphor of addiction when he said that "America is addicted to oil." Maybe it's more than just a metaphor. As a culture—and as individuals—we might truly be addicted to the products of our consumer system. In his book, *Consumed,* Benjamin Barber (2007) claims that an "indicator of the totalizing and homogenizing character of our consumer culture is its apparent addictiveness" (p.235). Barber suggests that addiction plays a major role in corporate consumer society. Powerfully addictive substances and behaviors, tobacco, alcohol, sugar- and fat-saturated foods, television and movies, video games, are major consumer items whose use is supported by an enormous amount of corporate advertising. In addition, the act of consumption itself has been turned into an addiction by corporate marketing, much of which is aimed at the most vulnerable segment of the population:

> Children once skipped rope, played house, hopped scotch, said Simon, sticked ball, and otherwise entertained themselves with more or less whatever the living environment conditioned by their imaginations could offer up. Today, play is commodity facilitated and consumer sponsored, a question of expensive gear, electronic video games, internet entertainment—all the right equipment regularly engineered in new and improved versions that demand constant repurchase and have the potential to induce addiction. Youthful consumers get hooked on the gear and the peripherals if not the games. (p. 238)

I would go one step further. It's not just consumption, but the whole of our consumer civilization that we are addicted to. And to see that addiction in this instance is not just a metaphor, consider how professionals diagnose substance dependence, the clinical term for addiction, defined as "[a] maladaptive pattern of substance use, leading to clinically significant impairment or distress [...]."[5] There are seven criteria for substance dependence listed in the DSM-IV-TR, the diagnostic manual used by mental health professionals. The first criterion is tolerance, the need for more of the substance to produce the desired effect, or a diminished effect with the same amount of the substance. We quickly develop tolerance for the material accessories of our consumer society. And it doesn't take long before we need more and bigger and faster. The average size of new homes, for example, has grown dramatically in the last few decades. And once we've actually attained more than we had previously, the positive psychological effects don't last long. Although life satisfaction and happiness can be affected by both positive and negative events, and can change over time, depending on a variety of factors (Diener, Lucas, & Scollon, 2006), most changes are transient, and people quickly revert to pre-event levels. In like manner, the purchase of expensive consumer goods can lead to a short-lived boost in satisfaction and mood. And then there is the often repeated platitude of the power hungry corporate executive who is never satisfied with any amount of wealth and power. Affluence and material wealth are easy to get used to, an effect referred to as the "hedonic treadmill" (Brickman & Campbell, 1971), and the desire for more is proportional to the amount that you already have: the more you have the more additional it takes in order to get the same psychological boost.

The second criterion is withdrawal, where the person experiences uncomfortable (in some cases life-threatening) physical symptoms when they stop taking the substance. Force an adolescent to give up his or her cell phone or favorite video game for a week, and you are sure to see symptoms of withdrawal that would rival a hard-core junkie going cold turkey. Pick any modern convenience and ask yourself how you would react if it were no longer available to you. Take even a relatively benign convenience such as the ability to enjoy a hot shower. As difficult as it might be for the typical American to imagine, most people on the planet have never in their life had a hot shower. In 1950, fewer than 30% of Americans had access to hot showers. As victims of hurricanes and earthquakes have found, the lack of running water is felt as something more than just a minor inconvenience, despite the fact that indoor plumbing is an extremely recent addition to the human experience. At the other end of the disaster spectrum, the corporate executive who takes his or her life in the wake of a failed business venture rather than face a reduction in wealth and power has become a cliché. And what about something as central to every facet of our civilized existence as oil? What would happen if petroleum were suddenly unavailable? Not only would most all transportation and industrial manufacture cease immediately, but we would also lose access to numerous products that are now literal necessities, many pharmaceuticals and plastic components used in the health-care industry, for instance, on which people's lives quite literally depend. The withdrawal symptoms of our oil addiction are in actual point of fact life threatening.

The remaining criteria for substance dependence are equally easy to apply to the products of modern civilization. The third criterion involves the person taking more of the substance than originally intended, or taking the substance for a longer period of time than originally planned. To see how this criterion applies, we need look no further than our high levels of personal debt, a reflection of a systemic inability to delay material gratification.[6] Many people have managed to rack up so much in the way of credit card and other personal debt that it is no exaggeration to say that they are living lives of indentured servitude to banking corporations. The fourth criterion involves the person wanting to quit or cut down but being unsuccessful at curtailing the use of the substance. Our inability to reduce carbon emissions fits here, as does our reluctance to transition away from gasoline powered automobiles. The fifth criterion is that the person spends a lot of personal time engaged in activities involved in obtaining the substance or in recovering from the effects of using. Shopping malls have become entertainment centers. Buying things, thinking about buying things, shopping in all of its forms represents the primary activity of many Americans—so much so that we no longer call ourselves citizens; we are *consumers*. The sixth criterion is that the person has given up or reduced participation in important activities as a result of substance use. What we have given up in our quest for ever-increasing consumption is no less than our freedom and our humanity. We spend more time interacting with things, with our consumer products, with our cell phones and our automobiles and our games and our entertainment centers and our computers, than with the important people in our lives.

And the seventh criterion—and this is a diagnostic clincher for many addiction counselors—the person continues to use the substance despite "knowledge of having a persistent or recurrent physical or psychological problem that is likely to have been caused or exacerbated by the substance." Our civilization is destroying the planet and us along with it. The reckless pursuit of our number one drug of choice has rendered a substantial proportion of the biosphere toxic to all life. Anthropogenic global climate change is more than likely already beyond anything we can do to reverse or even meaningfully ameliorate. Global warming deniers aside, we are very aware of the causes of the accelerating environmental degradation that is occurring all around us. Our industrial civilization is vacuuming the planet of all of its irreplaceable resources while the polar ice caps melt and species after species disappears forever, and yet we do nothing. It's business as usual. We drive. We shop. We consume. We wage wars that indiscriminately kill men women and children by the millions and litter the environment with depleted uranium munitions to protect our access to oil so that we can continue to drive, to shop, to consume. With respect to consumer civilization, we appear to meet *all* of the criteria for substance addiction. Moreover, it should be noted that for a person to be diagnosed with substance dependence according to the DSM, only three of the criteria need to be satisfied—any three. We clearly have a dependence problem.

Thinking about civilization as a kind of substance to which we have developed, both individually and collectively, a powerful addiction can explain the sometimes very strident reaction people have to suggestions that we need to make radical changes to the status quo. A similar reaction might be expected of a drug addict who discovers that his drug supply was going to be permanently cut off. Often, the drug, along with all of the activities that surround its procurement and use, has become such an intimate part of the addict's daily life that it is difficult to imagine life without it. The addictive substance becomes an organizing principle, providing an otherwise hollow life with purpose and meaning. The addiction, despite its clearly negative consequences, provides a level of comfortable certainty and predictability. To suggest life without the drug is to suggest a different life. And it's not just the drug itself, but everything associated with its use. When heroin addicts run short, many will stab themselves with empty needles, finding some relief in the ritual itself. In the same way, during an economic downturn many people find enjoyment in shopping for consumer items that they cannot afford to buy. Buying, consuming commercial products and services, working in exchange for money to spend on commercial products and services, acquiring an education in preparation for a career in order to work in order to be in a position to continue to buy commercial products and services are what gives life its meaning. We are consumers; it's what we do. To suggest that we need to put an end to the status quo and replace it with something radically different is to suggest that we need to acquire an entirely new sense of purpose, that we find entirely different ways of giving life its meanings, that we abandon our comfortable consumer chains, that we embrace the uncertainty of a life of freely chosen goals, that we become entirely different beings.

Viewing our problem as one of "civilization dependence" provides a way for us to understand why we continue to support our dehumanizing and unsustainable civilization even as its malignant tendrils wend increasingly deep into the tissue of the biosphere. The dependence metaphor also provides a potentially useful way of anticipating people's reactions to meaningful resistance. A psychologist by the name of James Prochaska and his colleagues (Prochaska, DiClemente, & Norcross, 1992) developed a theoretical model of the process of change, originally designed for use by addiction counselors. The model, called the *transtheoretical model of change,* has application beyond substance addiction to virtually any circumstance involving personal change. The theory views change as a recursive five stage process: *precontemplation, contemplation, preparation, action,* and *maintenance.* In the precontemplation stage, the thought of change is not being actively entertained. During this stage the person does not recognize that there is a problem to begin with. Or if a problem is acknowledged, it is seen as something minor, not serious enough to warrant taking action. This stage is marked by deep denial. For addicts at this stage, the positive benefits of the drug outweigh any perceived consequences, or if there are perceived consequences, they are not seen as directly relating to the use of the substance. For some addicts, the largest hurdle to cross is the transition from precontemplation to the next stage, the contemplation stage, to the realization that there is a problem and a need to take some action to change things. Unfortunately, just knowing and accepting that there is a problem and wanting to fix it is not sufficient to bring about change. And acquiring insight into the causes of a problem might be an illuminating experience, but insight by itself is not enough to fix the problem. What is missing at this stage is a meaningful commitment to change. During the preparation stage the person has committed to change and is in the process of putting together a strategic plan of attack. Acceptance of the severity of the problem is accompanied by a willingness to do whatever it takes to change. The action stage involves the actual enactment of the plan; this is where the "rubber meets the road" and the person begins to change his or her behavior. The final stage, the maintenance stage, is one of continued vigilance following successful change. The stages of this process are sequential and obligatory. A person can't jump directly from contemplation to action, for example. And regressive periods in which a person "relapses" back to previous stages are typical and to be expected. Also, the time course of each stage varies widely from person to person and situation to situation. Some people can spend years in the contemplation stage, where they know there is a problem that they need to address, and yet never advance to the preparation or action stages.

As a society, we are collectively in the precontemplation stage, both with respect to making changes necessary to deal with our addiction to global consumer culture and with respect to our civilization's impending disintegration. My hope is that readers who have made it this far in the book are at least to the contemplation stage. In the space the remains, I will explore issues related to the preparation and action stages. My discussion of these issues will be limited for a couple of reasons. First, I am in no position to know precisely what needs to be done to change our situation: an examination of potential courses of action would far exceed both my meager personal faculties and the limits of any single book. Second, any concrete suggestions I might make for specific acts of resistance could easily be interpreted as an attempt to incite violence, and make me, and this book, a target of the machine's protective defenses.

In the end, our addiction to civilization may not be the largest obstacle we will have to scale on the road to meaningful resistance. The "substance" of our consumer culture is not really all that powerful as an addictive agent; it gets whatever power it has over us by riding on the back of our evolutionary hard-wiring for an entirely different kind of lifestyle; the only way that our corporate consumer system succeeds is by continually exploiting our evolved psychological predilections. Despite the "resistance is futile" mantra, meaningful resistance might not be as difficult as we have been led to believe.

Freeing the Elephant

Raiding the Master's Toolshed

Our house is on fire and the arsonist is pointing a gun to our head and telling us to pour more gasoline. When compliance is the only option, it ceases to be compliance at all. Part of the power of the modern machine is that it eliminates alternatives. Accommodation to technology has left us acutely dependent. To resist the machine becomes, literally, to reject an entire way of being. Alternatives have been shut off from us—even the awareness that there are alternatives has been quashed. Practically speaking, there is no difference between an enduring ignorance of existing alternatives and having no alternatives to begin with. The progressive delusion, the myths of beneficence and inevitability, and a learned-helplessness-induced sense of inescapability provide the psychological mortar in the walls of the status quo. The machine of civilization is so omnipresent, the power complex so vast and convoluted, its protective garrisons so menacing, and our dependence so deeply embedded, that even if we acknowledge our captivity, and even if we manage a brief glimpse of the freedom that waits beyond our prison, thoughts of resistance are quickly sublimated, or relegated to passing fantasy, or expressed in passive-aggressive acts that have no lasting impact—token resistance.

Political activism is not the answer. Working within the system is not an option when the system itself is the problem. The problem is not in the specific ways in which the system is organized or in the particular configuration of policies, procedures, or laws. The problem is not the specific occupants of positions of authority. The problem is the existence of a system itself—any system. Any broad systematic organization of human activity runs counter to the human needs and interests of the individual people whose activity is being so organized. There is no political solution to the problem of civilization, and even the most enlightened progressive policies serve only to perpetuate the machine's agendas. If working to redirect the system from the inside is not an option, then how do we get at it from the outside when just recognizing that there is an outside, that the system is not simply the way the world works, is a challenge?

The power complex is multifaceted, with many interpenetrating levels. On the most local level, law enforcement provides a salient physical interface, frequently the machine's first line of defense against overt acts of resistance. Police are sanctioned—and equipped—to use overwhelming force. Engaging the mechanism of law enforcement, by staging violent protest or committing acts of vandalism or sabotage, for example, is unlikely to have any meaningful impact on the machine itself. Any act of resistance that triggers the machine's defenses is doomed to fail because of the machine's self-granted sanction—and capacity—to employ overwhelming force. There are far more bullets in the world than there are people. Attacking the most salient proximal source of control, local law enforcement, is an example of "targeting the fist." In a fist-fight, your opponent's knuckles are the most immediate source of the problem; nevertheless, to have any hope for success, you need to get around the fists and direct your attack toward the soft, vulnerable parts of your opponent's body. Confrontations between protestors and police in riot gear make for exciting news footage, but qualify only as token acts of resistance. Even those cases in which large-scale violent protest has led to regime change do not qualify as successful resistance against the machine of civilization. The power complex of the machine is composed entirely of replaceable parts—at all levels. In most cases regime change only facilitates the smooth operation of the machine. Oppressive dictators tend to be inefficient conduits of power and control.

Although the context of a post 9-11, Patriot Act world forces me to limit my discussion to abstract generalities, the shrewd reader should nonetheless be able to adduce the constituent ingredients of real resistance. Our enemy is a kind of machine, one that yet relies on human parts to function. In order to accomplish its reverse-adapted ends, the machine demands that we adopt modes of living that deviate dramatically from ancestral modes reflected in our genetic programming. There are limits to how far and how fast this mismatch can be extended; there are limits beyond which human beings start to lose their instrumental value to the machine. These limits have already been exceeded in several domains, as evidenced by the accelerating incidence of mental disorder and stress-related physiological illness. This fact represents a potentially exploitable limitation in the mechanical process: a targetable lynchpin in the system. Currently, the machine exploits the psychosocial deprivation caused by the mismatch by providing proxies and surrogates for what is absent. But these surrogates are ephemeral shadows that lack true potency to satisfy our needs.

One potential category of resistance, then, involves identifying and circumventing these proxies, while replacing them, inasmuch as is still possible, with the real thing. Mediated communication makes for a good example case. Communication technology has become reverse adapted such that our innate needs for affiliation and association are distorted and channeled to support the continued expansion of the global communication industry. To the extent that we still have conversations, they are frequently reduced to a loosely connected aggregate of sound-bites or a spattering of inarticulate screen text. The resulting affiliations and associations are impoverished two-dimensional affairs that groom us for shallow transient human relationships and a tolerance for isolation—both of which increase our instrumentality for the machine. A potential line of resistance here might involve attacks on the physical and organizational infrastructure supporting mediated communication while simultaneously providing the opportunity for meaningful face-to-face association. I will leave specific details to the reader's creative imagination.

What other domains involve satisfying inborn human needs with surrogates? The educational system seems an obvious candidate. The realm of formal education is a collection of proxies directed at a variety of psychological needs, including esteem, belongingness, and affiliation, among others. What happens if formal education is replaced with actual opportunities to learn from each other? What happens to the individual? What happens to society as a whole? Or, more importantly, what happens to the machine if one of the primary mechanisms for the fine partitioning and isolation of knowledge is taken off line? Without a standardized formal educational system culminating in the powerful research university, we don't have specialists in atomic physics or nanotechnology, we don't have experts in propaganda and other forms of psychological manipulation. Our educational system itself is merely one example of how the machine employs systematic psychological manipulation. As with mediated communication, resisting the educational system would come down to some combination of attacks on the physical and organizational infrastructure combined with the provision of alternatives ways of satisfying the human needs currently being satisfied in an ersatz way by the system, alternatives more in line with our authentic human nature.

Local self-reliance represents another potential category of resistance. The machine's power is reduced in exact proportion to which local communities are able to provide for their own needs. Permaculture and related approaches to community living involve exactly the two-pronged approach to resistance that I suggested above: weaken the mechanical infrastructure while simultaneously offering more authentic human alternatives. One of the machine's primary sources of control is dependency. Local self-reliance is by definition a reduction in dependency. The local provision of physical necessities such as food and shelter also entails a high degree of cooperation and mutual goal pursuit among community members, and thus establishes preconditions for satisfying many of the less tangible psychological needs for affiliation that are being exploited by the machine. And even an urban community garden is a small but real act of resistance.

There are several potential beneficial side-effects of community self-reliance. First, it provides potential opportunities to immunize people against learned helplessness. It also provides circumstances that have the potential to enhance self-regulation. Both of these serve to loosen the machine's grip in its remaining domains of control. In addition, community cohesion and the attention to the natural environment necessitated by the local provision of physical necessities are conditions that counter arrested development. Arrested development is a product of blocked or distorted epigenetic programs, programs that are designed for the kind of input provided by close contact with nature and life-ways embedded in natural, multigenerational social groups. Although, we may not have enough time to allow for the passing of even a single generation before we stop the machine of civilization. Resistance must be more direct and immediate if we are to redeem enough of the natural world for future generations to have anything resembling a truly human experience.

Local self-reliance involves a reduction in centralized power. Widespread local self-reliance would reduce the machine from a dominating force to a minor annoyance. Decentralization in general appears to work on at least two levels. First, and most obvious, it reduces the machine's power. Second, decentralization, and the concomitant elevation of the importance of the local community, feeds into our human nature as a species evolved for life in small social groups. It is important to note, however, that decentralization would have to be enacted in multiple disparate domains. The power complex itself is a collection of largely self-contained systems of control, each capable of carrying out its function in the absence of any overarching authority. There is no single target at the top, no central node that if removed would render the entire system inoperative. Rather, we are dealing with a large number of interconnected and loosely interdependent but nonetheless modular systems. The sudden elimination of central government, for example, would not dismember the military complex or automatically render local law enforcement inert. These systems are designed to function as integrated wholes with the minimum necessary external input.

Working toward local food self-reliance may be a particularly potent form of resistance. The May 2009 issue of *Scientific American Magazine* included an article entitled "Could Food Shortages Bring Down Civilization?" The author,

Lester Brown, made a cogent case for an affirmative answer to that question. Rising demand in combination with eroding soils, environmental changes brought on by global warming, and the depletion of non-rechargeable "fossil" aquifers means that widespread food shortages are a virtual certainty in the very near future. When this happens, already unstable third-world governments will collapse like dominos and the global economy will disintegrate, leading to a worldwide crisis that could unravel the threads of modern civilization. After some considerable arm-waving about how we need to make "a monumental shift away from business as usual," Brown offers technical solutions that are sweeping in terms of the policy changes involved, but that essentially leave the status quo intact. He includes such things as cutting carbon emissions by 80%, stabilizing global population at eight billion, magically eradicating poverty, and reversing the destruction of forests, soil, and aquifers.

There are at least two potentially erroneous implicit assumptions woven into Brown's discussion. One is that the disintegration of civilization is necessarily a bad thing, something we should prevent at all cost. It is an open question as to whether any single facet of modern civilization is truly worth preserving to begin with—an open question that is very likely to have a negative answer when all is said and done. Not only is the disintegration of civilization ultimately unavoidable, but it is something that we should actively encourage, something that we should facilitate, something that we should direct our effort to bring about in a way that does the least amount of long-term damage. Looming political and economic crises generated by food shortages might just be the catalyst we need to redirect our collective attention toward the humane dismemberment of the global corporate system (see Rockerfeller, 2009). Another implicit assumption that Brown makes is that the problems that he outlines are not in fact inevitable consequences of civilization itself. Civilization is the root cause of aquifer depletion, overpopulation, poverty, soil erosion, global warming, and forest destruction. It's not how we do civilization; it's civilization itself. It's not just our modern civilization. Civilization is destructive by definition.

All civilizations have generated these kinds of problems. The difference between past and present civilization is merely one of scope. The destruction caused by the great civilizations of the past was (relatively) more localized. The ancient Egyptian civilization, the Mayan, the ancient Greek and Roman civilizations were not global in their reach. And, of course, they were not industrialized; they were powered largely by human labor, an energy resource with a far smaller environmental footprint than fossil fuels. The point is that the serious problems that Brown claims will eventually lead to the disintegration of civilization are themselves natural byproducts of civilization. But Brown is right about one thing: food is the keystone of civilization. It has been the keystone and thus the Achilles heel of every civilization. As such, food production and distribution are targetable linchpins. Local food self-reliance then becomes, not just a potential way of accelerating the collapse of the corporate food machine, but a prophylactic for surviving at least the initial stages of civilization collapse.

Anarcho-Primitivism as a Trajectory

In terms of philosophical-political perspective, the evolutionary psychology-informed critique of civilization that I have been offering fits loosely under the label *anarcho-primitivism.* Anarcho-primitivist thought comes in a variety of flavors. Some varieties focus more on primitivism, and emphasize the negative impact of industrial technology and the human benefits of a return to a technological state better aligned with our evolutionary roots. Some focus more on anarchism and the need to extract ourselves from hugely oppressive systems of power and control that actively prevent the free expression of our innate human nature. Some focus on the natural world and the need to establish ecologically-sensitive lifestyles and harmonious relationships with the biosphere (a.k.a. green anarchism). Regardless of the primary focus, the variants of anarcho-primitivism share at least this in common: that global industrial civilization is a bad idea and needs to stop—very soon. And if we cannot put an immediate end to industrial civilization, let's at least work toward rendering it irrelevant; let's establish the preconditions for its extinction; let's forge a monkey wrench large enough to terminally jam the gears. And if we can bring down the global machine in a way that limits the human and ecological toll, so much the better.

In the late 1970s, a Japanese farmer by the name of Mansanobu Fukuoka wrote *The One Straw Revolution.* Fukuoka's book—really a manifesto—presents an approach to organic farming that can serve as a powerful model for a commonsense approach to living in general. He calls his method "do-nothing" farming. It is based on the premise that working with the land's evolved natural propensities can ultimately yield far superior results compared to modern farming with its monoculture and its labor-intensive environmentally destructive techniques. Modern industrial farming attempts to force nature, or impose an artificial structure on the natural world. Fields are plowed and planted with crops that need to be fertilized because the soil's ability to sustain growth has been destroyed by the cultivation itself. Herbicides are then applied to keep the "weeds" at bay. All of this requires an enormous amount of human and natural resources. Fukuoka's do-nothing approach is simply to scatter seed on an existing uncultivated field. Along with the desired crop, "weeds" of a certain type are planted to keep other weeds in check. The straw from one harvest is allowed to sit on the field and decompose naturally even as the next season's crop is being sown. After a few seasons, the field is producing almost as much as a commercially cultivated and chemically treated field—but without either the cultivation or the chemicals. The plants are healthier, and there is a net improvement in the soil season by season. Even poor land and depleted soil can be resurrected by his methods.

Fukuoka's do-nothing approach to farming has something important to offer us here, something more than mere metaphor. Industrial civilization forces us to live in an unnatural, highly "cultivated" manner, and by living in this way we destroy our environment in the same way that plants forced to live in industrial monoculture exhaust the soil. And, as with the crops of industrial agriculture, it takes an enormous amount of energy and resources to maintain our lifestyle because we are being forced to live in conditions that run counter to our evolved propensities. Fukuoka's solution is to stop the machines, let the soil and the plants do what they have been designed to do through several hundred million years of evolutionary fine-tuning. Likewise, the solution to restoring our social environment is to stop the machine of civilization, stop forcing our lives into conformity with an artificial and inhuman mode of being. Out of civilization's remains will eventually emerge fertile ecological and social "soil" for nurturing all of our human needs. The problem will be one of stopping the cultivators, putting an end to the mechanical disturbance, and then having the patience to allow the dust to settle—and the fortitude to accept that the dust of civilization's collapse will still be in the air that our great-great grandchildren breathe. So the implementation of Fukuoka's solution breaks down to two parts. First, we need to put an end to the mechanical cultivation. We need to stop the industrial machine that is devouring the natural world and degrading our humanity. Second, in its place we need to cultivate patience, we need to allow time for the "soil" to heal itself, to reacquire its ability to sustain and nourish. The first part will be the hard part. Perhaps impossibly hard. But we will be supported in the second part by our own evolved human nature.

In *Ishmal,* Daniel Quinn's gorilla talks about the voice of mother culture, by which he means the memes and narratives of civilization that we have incorporated so deeply into our worldview that we are blind to their organizing and biasing presence. One of the most prevalent unspoken and unquestioned truths of civilization is the need for organizational hierarchy. Civilization is built on division of labor, specialization, expertise, the unequal access to resources, and the power differentials that result from these. It would not be hyperbole to equate civilization with these things. The need for unequal distribution of power and control is a cornerstone of the civilized worldview. The understanding that decision making needs to be grounded in hierarchy is woven into the very fiber of our "democratic" system of government. Anarchy, one half of the anarcho-primitivist formula, is the barefaced denial of the legitimacy of hierarchy. The term *anarchy* is frequently used as a synonym for chaos. The conflation of anarchy with chaos comes right out of mother culture's insistence that power must be unequally distributed, and that without top-down control all would be confusion. Anarchy in this sense reflects the assumed state of disorder that would result from a lack of control over the masses. But who or what it is that should have this control in the first place, and what makes the exercise of this control legitimate, is rarely mentioned; and when it is, it is through the use of abstract and reified terms such as *the social contract, the government, the rule of law*.

In addition to chaos, anarchy is frequently coupled with violence: the caricature of the anarchist as a bomb-carrying thug. Violence, unfortunately, is an unavoidable element of the anarchy equation. But anarchy is not the source of the violence. Violence is built into our hierarchical system. The only way to maintain a system with such dramatic disparities in power and access to resources is through violence. And the more extreme the disparities, the more violent the methods of maintenance need to be. Because violence is the mortar that holds the bricks of the hierarchical system in place, any meaningful attempt to dismantle the system will elicit violence. Also, because anarchists are usually people who want to overthrow the existing power structure, it makes some sense to think that they would probably employ violent means to do so. How could it be otherwise? You need to fight fire with fire. But here we need to distinguish between ends and means, between anarchy as a goal-state, and the methods for bringing that state about. And it is important to keep in mind that when it comes to means there is no necessary relationship between violence and effectiveness. It may be possible to bring our corporate consumer system down in a relatively nonviolent fashion simply by finding a way for enough of us to avoid playing the corporate consumer game.

Then again, we may have to blow a few things up.

Anarcho-primitivism is redundant as a hyphenated two-term label. True anarchy also has to be primitive. Anything else is anarchy of only a limited sort, restricted to the shared control over the means of production, for instance. Production implies division of labor which in turn implies hierarchy. And civilization implies both. So traditional anarchist perspectives, views that focus on the need to for a more collective form of government, or the need for a more equitable distribution of resources, for example, are not anarchy in the anarcho-primitivist sense. Anarchy, in its anti-hierarchy sense, is a necessary guiding principle for dismantling civilization. Anything less, say, democratic socialism, allows for preservation of the hierarchical status quo. Hierarchy is oppressive whether or not we are allowed to choose our oppressors. True anarchy, then, is a return to the primitive social organization of our subsistence ancestry. Likewise, true primitivism also has to be anarchism. A return to our subsistence roots implies lifestyles that involve only a very limited division of labor tied to individual differences in natural ability, personal proclivity, and local custom.

The irrational fear of anarchy that seems so prevalent can be tied to the conflation of anarchy with chaos, and reflects a successful strategy by mother culture to keep us in line. But there is a deep-seeded psychological component to this as well that reflects a fear of personal freedom; true freedom is a condition with which we have little personal experience. Temple Grandin (1995) is famous for her work with cows. She is also famous for being diagnosed with Asperger's syndrome, a variant of autism in which much of the person's higher intellectual functions appear to be spared. She claims that her Asperger's gives her the ability to get into the cows' mind-space and understand, for example, how a cow feels when it is undergoing the stressful transition from pen to slaughterhouse. She reports having a pivotal insight while watching calves being inoculated on her Aunt's Arizona ranch. The ranchers used something called a squeeze chute, a device that literally clamps the calf tightly from the sides so that it is unable to move while it receives its dose of antibiotics and growth hormones. What she noticed—an apparent paradox—was that many of the animals calmed down and relaxed when they were being constrained. This insight eventually led to the redesigning of various physical structures used with cattle and other factory farmed animals in slaughterhouses and dairies around the world.

That there is comfort in constraint suggests that freedom causes anxiety. To be truly free to pursue your own freely chosen goals means that you are responsible for the outcome of your goal pursuit. That can be a very scary thing. It is also a very rare thing. Most of us are pursuing goals that have been created for us, goals that we would never choose to pursue if we were given an actual choice. What kind of person would freely submit to a 40-plus hour work week in pursuit of fleeting material wealth and the dubious promise of a "better" future? Who would choose to submit to a state-(and corporate)-defined formal education designed primarily to instill the skills and habits of mind necessary to become an effective consumer? Who would willingly renounce his or her natural rights to clean water, fresh air, and a healthy landbase?

The discomfort of freedom is something that most of us eagerly trade for the illusion of safety found in artificially-crafted constraint; we gladly give control of our lives to other people and things so that we don't have to bear the existential burden of freedom. The life of a wage-slave is preferred to the life of a free (wo)man. The pursuit of convenience, comfort, and mindless entertainment is preferred to a freely chosen life of purpose. Identity foreclosure and hollow imitation is preferred to a life-long journey of self-discovery. We have become like timid cattle—both of us animals that bear little resemblance to our spirited and fearsome evolutionary ancestors—living in domestic servitude to powers of which we choose to have little awareness and even less understanding, held comfortably placid in our corporate-consumer squeeze chutes.

But our squeeze chutes are mythical constructions that provide as much actual restraint as a butterfly net has over an elephant.

On the terms imposed by technocratic society, there is no hope for mankind except by 'going with' its plans for accelerated technological progress, even though man's vital organs will all be cannibalized in order to prolong the megamachine's meaningless existence. But for those of us who have thrown off the myth of the machine, the next move is ours: for the gates of the technocratic prison will open automatically, despite their rusty hinges, as soon as we choose to walk out (Mumford, 1970, p.435).

Choosing to walk out is not as simple as it sounds, of course. First we need to reclaim our faculties of choice. First we need to relinquish our role as servomechanism and extract ourselves from the pursuit of goals that have been reverse-adapted to accommodate the needs of the machine at the expense of our humanity. First we need to reject the legitimacy of all forms of hierarchical social-political arrangements. Hierarchy is a butterfly net. Ultimately, the power exercised by the top levels of hierarchy is generated from below, fueled by belief in its legitimacy, and maintained by a misguided identification with the system, an illusory sense that the machine is acting in the service of our freely chosen ends, and a failure to perceive the mechanisms by which our goals and actions are being directed:

> [O]bedience to authority requires each of us to first participate in the myth-making process of creating authority figures who then must legitimize their authority through the evidence of our submission and obedience to them [and] the reason we can be manipulated so readily is precisely because we maintain an illusion of personal invulnerability and personal control, all the time being insensitive to the power of social forces... (Zimbardo, 1974, p. 566)

I have argued that the major problems of civilization can be traced to a mismatch between our evolved predilections and the demands of a mechanically organized existence. We emerge from the womb prepared to mature into life-ways deeply embedded in natural systems tied to a specific geographical place, and into a meaning-infused social world based on small life-long groups organized around relatively egalitarian participation. What we find instead is a lifestyle based on estrangement from nature, transient social relationships, large-scale oppression, and superficial material gratification.

But the machine of civilization can yet only operate within the limits imposed by our genetic design; its mechanisms yet need to accommodate human tolerances. And the mismatch may prove to be the source of our eventual salvation: every child is born expecting the Pleistocene.

Notes

[1] There are some superficially non-mechanical ways of conceptualizing facets of civilization, as complex nonlinear systems, for example. Some may wish to argue that civilizations, societies in general, are organisms, not machines. But we think about organic nature almost entirely in terms of mechanism. Blood is *pumped* by the heart and *transported* via the circulatory *system*. DNA is a kind of blueprint. Mitochondria are responsible for energy production. Even the complex dynamic oscillations of populations are driven by various mechanisms. So to call society an organism is simply to disguise the machine behind a second-order mechanical metaphor.

[2] http://www.socialstudies.org/standards/strands#1

[3] Another problem with the human bloodlust hypothesis is that it doesn't account for the lag time between human presence and extinction. "Importantly, the dates suggest that the local decline in biological diversity was initiated ~75,000 years before the colonisation of humans on the continent [of Australia]. Collectively, the data are most parsimoniously consistent with a pre-human climate change model for local habitat change and megafauna extinction, but not with a nearly simultaneous extinction of megafauna as required by the human-induced blitzkrieg extinction hypothesis" (Price et al., 2011, p. 10). And from Barnosky & Lindsey (2010, p. 10): "...on a continental scale most megafauna have last appearances after human arrival, but seem to last at least 1000years after first human presence. Some taxa apparently survived >6000years after humans entered South America and >1000years after the end-Pleistocene climatic changes. Last-appearance patterns for megafauna differ from region to region, but in Patagonia, the Argentine and Uruguayan Pampas, and Brazil, extinctions seem more common after humans arrive and during intensified climatic change between 11.2 and 13.5ka. This pattern suggests that a synergy of human impacts and rapid climate change—analogous to what is happening today—may enhance extinction probability. Nevertheless, even in these regions, some megafauna persisted for thousands of years after human arrival and after the climate warmed." Note, for the record, that the rapid climate change happening today is itself a human impact, so the "synergy of human impacts and rapid climate change" happening today is not a synergy at all.

[4] Note that the childish projection of agency onto inanimate features of the environment is different from the perception of a shared universal awareness or the perception of psychological qualities in natural phenomena found in many hunter-gatherer cultures. The former is an egocentric overextension, the latter is a transcendent awareness.

[5] DSM-IV-TR, p. 197.

[6] But see Graeber (2011) for a discussion of personal debt as a systemic feature of civilization and a potent tool for oppression and control.

References

Abbott, A. (2011). City living marks the brain. *Nature*, 474(7352), 429.

Abrutyn, S. & Lawrence, K. (2010). From chiefdom to state: Toward an integrative theory of the evolution of polity. *Sociological Perspectives*, 53(3), 419-442.

Adams, W. W. (2005). Ecopsychology and phenomenology, toward a collaborative engagement. Existential Analysis, 16, 269-283.

Anderson, C. J. (2003). The psychology of doing nothing: Forms of decision avoidance result from reason and emotion. *Psychological Bulletin*, 129(1), 139-166.

Asch, S. E. (1956). Studies of independence and conformity: A minority of one against a unanimous majority. *Psychological Monographs,* 70, (9, Whole No. 416).

Baillargeon, R. (1987). Young infants' reasoning about the physical and spatial properties of a hidden object. *Cognitive Development, 2,* 179-200.

Barber, B. R. (2007). *Consumed.* New York: Norton.

Barnosky, A. D., & Lindsey, E. L. (2010). Timing of Quaternary megafaunal extinction in South America in relation to human arrival and climate change. *Quaternary International, 217(1/2),* 10-29.

Bereczkei, T. (2000). Evolutionary psychology: A new perspective in the behavioral sciences. *European Psychologist*, 5(3), 175-190.

Berry, T. (1988). *The Dream of the Earth.* San Francisco: Sierra Club Books.

Bird, D. W., & O'Connell, J. F. (2006). Behavioral ecology and archaeology. *Journal of Archaeological Research, 14,* 143-188.

Bishop, B. R., & Beckman, L. (1971). Developmental conformity. *Developmental Psychology*, 5(3), 536.

Brickman, P. & Campbell, D. T (1971). Hedonic relativism and planning the good society. In M.H. Appley (Ed.), *Adaption Level Theory: a Symposium.* New York: Academic Press.

Brown, L. R. (2009). Could food shortages bring down civilization? *Scientific American,* 300(5), 50-57.

Cacioppo, J. T., Fowler, J. H., & Christakis, N. A. (2009). Alone in the crowd: The structure and spread of loneliness in a large social network. *Journal of Personality and Social Psychology*, 97(6), 977-991.

Carr, N. (2010). *The Shallows.* New York: Norton.

Conway, L., & Schaller, M. (2005). When authorities' commands backfire: Attributions about consensus and effects on deviant decision making. *Journal of Personality and Social Psychology*, 89(3), 311-326.

Cosmides, L, Tooby, J., & Barkow, J. H. (1992). Introduction: Evolutionary psychology and conceptual integration. In J. H. Barkaow,   L. Cosmides & J. Tooby (Eds.). *The Adapted Mind: Evolutionary*

Psychology and the Generation of Culture. New York: Oxford University Press.

Davis, K. M., & Atkins, S. S. (2004). Creating and teaching a course in ecotherapy: We went to the woods. *Journal of Humanistic Counseling, Education & Development,* 43(2), 211-218.

Diener, E., Lucas, R. E., & Scollon, C. (2006). Beyond the hedonic treadmill: Revising the adaptation theory of well-being. *American Psychologist,* 61(4), 305-314.

Diamond, J. (1992). *The Third Chimpanzee.* New York: HarperCollins.

Diamond, S. (1974). *In Search of the Primitive.* New Brunswick: Transaction.

Doherty-Sneddon, G., Anderson, A., O'Malley, C., Langton, S., Garrod, S., & Bruce, V. (1997). Face-to-face and video-mediated communication: A comparison of dialogue structure and task performance. *Journal of Experimental Psychology: Applied,* 3(2), 105-125.

Duffy, J. (2000). Never hold a pencil. *Written Communication,* 17(2), 224.

Dunbar, R. I. M. (1993). Coevolution of neocortical size, group size, and language in humans. *Behavioral and Brain Sciences, 16,* 681-735.

Eidelman, S., Crandall, C. S., & Pattershall, J. (2009). The existence bias. *Journal of Personality and Social Psychology,* 97(5), 765-775.

Fast, N. J., Gruenfeld, D., Sivanathan, N., & Galinsky, A. D. (2009). Illusory control: A generative force behind power's far-reaching effects. *Psychological Science,* 20(4), 502-508.

Fitzsimons, G. M., & Fishbach, A. (2010). Shifting closeness: Interpersonal effects of personal goal progress. *Journal of Personality and Social Psychology,* 98(4), 535-549.

Foster, K. R., & Kokko, H. (2009). The evolution of superstitious and superstition-like behaviour. *Proceedings of the Royal Society B: Biological Sciences,* 276(1654), 31-37.

Frassetto, L. A., Schloetter, M., Mietus-Snyder, M., Morris, R. C., & Sebastian, A. (2009). Metabolic and physiological improvements from consuming a Paleolithic, hunter-gatherer type diet. *European Journal of Clinical Nutrition, 63,* 947=955.

Frumkin, H. (2001). Beyond toxicity: Human health and the natural environment. *American Journal of Preventative Medicine, 20,* 234-240.

Fukuoka, M. (1978). *The One-Straw Revolution.* New York: New York Review of Books.

Gardiner, E., & Jackson, C. J. (2010). Eye color Predicts Disagreeableness in North Europeans: Support in Favor of Frost (2006). *Current Psychology,* 29(1), 1-9.

Gibson, J. J. (1986). *The Ecological Approach to Visual Perception.* London: Lawrence Earlbaum Associates.

Glendinning, C. (1994). *"My Name is Chellis & I am Recovering from Western Civilization."* Shambhala

Gould, S. J., & Eldredge, N. (1977). Punctuated equilibria: The tempo and mode of evolution reconsidered. *Paleobiology, 3,* 115-151.

Graeber, D. (2011). *Debt: The First 5000 Years.* Melville House.

Grandin, T. (1995). *Thinking in Pictures: And other Reports from My Life with Autism.* New York: Doubleday.

Greeno, J. G. (1994). Gibson's affordances. *Psychological Review,* 101(2), 336-342.

Gruenfeld, D. H., Inesi, M., Magee, J. C., & Galinsky, A. D. (2008). Power and the objectification of social targets. *Journal of Personality and Social Psychology,* 95(1), 111-127.

Gullone, E. (2000). The biophilia hypothesis and life in the 21st century: Increasing mental health or increasing pathology? *Journal of Happiness Studies, 1,* 293 – 321.

Hardin, G. (1968). The tragedy of the commons. *Science, 162,* 1243 – 1248.

Heise, D., Lenski, G., & Wardwell, J. (1976). Further notes on technology and the moral order. *Social Forces,* 55(2), 316-337.

Hill, R. A. & Dunbar, R.I.M. (2003). Social network size in humans. *Human Nature, 14,* 53-72.

Holzhaider, J. C., Hunt, G. R., & Gray, R. D. (2010). The development of pandanus tool manufacture in wild New Caledonian crows. *Behaviour,* 147(5/6), 553-586.

Hughes, C., & Ensor, R. (2007). Executive function and theory of mind: Predictive relations from ages 2 to 4. *Developmental Psychology,* 43(6), 1447-1459.

Hume, D. (1975). An enquiry concerning human understanding. In L. A. Selby-Bigge (Ed.), *Enquiries Concerning Human Understanding and Concerning Principles of Morals.* Oxford: Clarendon Press. (Original published 1748)

Hume, D. (1992). *A treatise on Human Nature.* New York: Prometheus. (Original published 1739)

Indurkhya, B. (2007). Rationality and reasoning with metaphors. *New Ideas in Psychology,* 25(1), 16-36.

Jerison, H. (1973). *Evolution of the Brain and Intelligence.* Academic Press: New York.

Kahn, P. H., Friedman, B., Gill, B., Hagman, J., Severson, R. L., Freier, N. G., & ... Stolyar, A. (2008). A plasma display window?--The shifting baseline problem in a technologically mediated natural world. *Journal of Environmental Psychology,* 28(2), 192-199.

Kahn, P.H., Severson, R.L., & Ruckert, J.H. (2009). The human relationship with nature and technological nature. *Current Directions in Psychological Science, 18,* 37-42.

Kalpidou, M., Costin, D., & Morris, J. (2011). The relationship between facebook and the well-being of undergraduate college students. *CyberPsychology, Behavior & Social Networking,* 14(4), 183-189.

Kaczynski, T. J. (2010). *Technological Slavery.* Port Townsend, WA: Feral House.

Kellert, S. R. (1997). *Kinship to Mastery: Biophilia in Human Evolution and Development.* Washington D.C.: Island Press.

Kim, E. (2003). Horticultural therapy. *Journal of Consumer Health on the Internet,* 7(3), 71.

Kruger, J., Epley, N., Parker, J., & Ng, Z. (2005). Egocentrism over e-mail: Can we communicate as well as we think? *Journal of Personality and Social Psychology,* 89(6), 925-936.

Kuo, F. E, & Taylor, A.F. (2004). A potential natural treatment for attention deficit/hyperactivity disorder: Evidence from a national study. *American Journal of Public Health, 94,* 1580-1586.

Lakoff, G. & Johnson, M. (1980). *Metaphors We Live By.* Chicago: University of Chicago Press.

Landau, M. J., Meier, B. P., & Keefer, L. A. (2010). A metaphor-enriched social cognition. *Psychological Bulletin,* 136(6), 1045-1067.

Lau, R. R., & Schlesinger, M. (2005). Policy frames, metaphorical reasoning, and support for public policies. *Political Psychology,* 26(1), 77-114.

Levy, D. (1997). *Tools of Critical Thinking: Metathoughts for Psychology.* Boston: Allyn & Bacon.

Lewandowsky, S., Mundy, M., & Tan, G. A. (2000). The dynamics of trust: Comparing humans to automation. *Journal of Experimental Psychology: Applied,* 6(2), 104-123.

Lickliter, R. & Honeycutt, H. (2003). Developmental dynamics: Toward a biologically plausible evolutionary psychology. *Psychological Bulletin, 129,* 819-835.

Louv, R. (2006). *The Last Child in the Woods: Saving our Children from Nature-Deficit Disorder.* Chapel Hill: Algonquin

MacDonald, K. B. (2008). Effortful control, explicit processing, and the regulation of human evolved predispositions. *Psychological Review,* 115(4), 1012-1031.

MacKenzie, D. (2008). The end of civilisation. *New Scientist,* 197(2650), 28-31.

Maier, S. F., & Seligman, M. E. P. (1976). Learned helplessness: Theory and evidence. *Journal of Experimental Psychology: General, 105,* 3-46.

Marks, D. F. (2007). Literacy not intelligence moderates the relationships between economic development, income inequality and health. *British Journal of Health Psychology,* 12(2), 179-184.

McNeil, B. J., Pauker, S. G., Sox, H. C., & Tversky, A. (1982). On the elicitation of preferences for alternative therapies. *The New England Journal of Medicine, 306,* 1259–1262.

McPherson, M. Smith-Lovin, L., & Brashears. M. E. (2006). Social isolation in America: Changes in core discussion networks over two decades. *American Sociological Review, 71,* 353-375.

Metzner, R. (1999). *Green Psychology: Transforming Our Relationship to the Earth.* Vermont: Park Street Press.

Michalski, R. L., & Shackelford, T. K. (2010). Evolutionary personality psychology: Reconciling human nature and individual differences. *Personality & Individual Differences*, 48(5), 509-516.

Mikulincer, M. (1986). Attributional processes in the learned helplessness paradigm: Behavioral effects of global attributions. *Journal of Personality and Social Psychology*, 51(6), 1248-1256.

Milgram, S. (1963). Behavioral study of obedience. *Journal of Abnormal and Social Psychology, 67,* 371-378.

Mohawk, J. (1992) In search of noble ancestors. In C.W. Galley (Ed.) *Civilization in Crisis: Anthropological Perspectives.* Gainseville: University Press of Florida.

Moors, A., & De Houwer, J. (2005). Automatic Processing of Dominance and Submissiveness. *Experimental Psychology*, 52(4), 296-302.

Moretti, F., de Ronchi, D., Bernabei, V., Marchetti, L., Ferrari, B., Forlani, C. & ... Attli, A. (2011). Pet therapy in elderly patients with mental illness. *Psychogeriatrics*, 11(2), 125-129.

Mumford, L. (1934). *Technics & Civilization.* Chicago: University of Chicago Press.

Mumford, L. (1966). *Technics and Human Development.* New York: Harcourt Brace Jovanovich.

Mumford, L. (1970). The Myth of the Machine: The Pentagon of Power. New York: Harcourt.

Muraven, M., & Baumeister, R. F. (2000). Self-regulation and depletion of limited resources: Does self-control resemble a muscle? *Psychological Bulletin*, 126(2), 247-259.

Nairne, J. S., Pandeirada, J. S., Gregory, K. J., & Van Arsdall, J. E. (2009). Adaptive memory: Fitness relevance and the hunter-gatherer mind. *Psychological Science)*, 20, 740-746.

Neuman, W., & Pollack, A. (2010, May 3). Farmers cope with Roundup-resistant weeds. *The New York Times.* Retrieved from http://www.nytimes.com

O'Dea, K., Spargo, R., & Akerman, K. (1980). The effect of transition from traditional to urban life-style on the insulin secretory response in Australian Aborigines. *Diabetes Care*, 3(1), 31-37.

O'Dea, K., White, N., & Sinclair, A. (1988). An investigation of nutrition-related risk factors in an isolated Aboriginal community in northern Australia: advantages of a traditionally-orientated life-style. *The Medical Journal of Australia, 148* (4), 177-80.

Öhman, A, Mineka, S. (2001). Fears, phobias, and preparedness: Toward an evolved module for fear and fear learning. *Psychological Review, 108,* 483-522.

Orians, G.H. & Heerwagen, J.H. (1992). Evolved responses to landscapes. In L. Cosmides & J. Tooby (Eds.). *The Adapted Mind: Evolutionary Psychology and the Generation of Culture.* New York: Oxford University Press.

Ornstein, R. & Ehrlich, P. (1989). *New World, New Mind.* New York: Simon & Schuster.

Osiurak, F., Jarry, C., & Le Gall, D. (2010). Grasping the affordances, understanding the reasoning: Toward a dialectical theory of human tool use. *Psychological Review*, 117(2), 517-540.

Peen, J., & Dekkar, J. (2004). Is urbanicity an environmental risk-factor for psychiatric disorders? *The Lancet, 363*, 2012-2013.

Piaget, J. (1954). *The Construction of Reality in the Child.* New York: Basic Books.

Pinker, S. (1994). *The Language Instinct.* New York: William Morrow and Company.

Platt, J. (1973). Social Traps. *American Psychologist*, 641-651.

Pollet, T. V., Roberts, S. B., & Dunbar, R. M. (2011). Use of social network sites and instant messaging does not lead to increased offline social network size, or to emotionally closer relationships with offline network members. *CyberPsychology, Behavior & Social Networking*, 14(4), 253-258.

Postman, N. (1992). *Technopoly.* New York: Knopf.

Postmes, T., Spears, R., & Lea, M. (2002). Intergroup differentiation in computer-mediated communication: Effects of depersonalization. *Group Dynamics: Theory, Research, and Practice*, 6(1), 3-16.

Price, G. J., Webb, G. E., Zhao, J., Feng, Y., Murray, A. S., Cooke, B. N., & ... Sobbe, I. H. (2011). Dating megafaunal extinction on the Pleistocene Darling Downs, eastern Australia: the promise and pitfalls of dating as a test of extinction hypotheses. *Quaternary Science Reviews, 30(7/8),* 899-914.

Prislin, R., Sawicki, V., & Williams, K. (2011). New majorities' abuse of power: Effects of perceived control and social support. *Group Processes & Intergroup Relations*, 14(4), 489-504.

Prochaska, J. O., DiClemente, C. C., & Norcross, J. C. (1992). In search of how people change: Applications to addictive behavior. *American Psychologist, 47*, 1102-1114.

Pushkina, D., & Raia, P. (2008). Human influence on distribution and extinctions of the late Pleistocene Eurasian megafauna. Journal Of Human Evolution, 54(6), 769-782.

Quinn, D. (1992) *Ishmael.* New York: Bantum.

Rafanell, I., & Gorringe, H. (2010). Consenting to domination? Theorising power, agency and embodiment with reference to caste. *Sociological Review*, 58(4), 604-622.

Ramírez, E., Maldonado, A., & Martos, R. (1992). Attributions modulate immunization against learned helplessness in humans. *Journal of Personality and Social Psychology*, 62(1), 139-146.

Rasmussen, M, Guo, X., Wang, Y., Lohmueller, K. E. Rasmussen, S., Albrechtsen, A.,...Willerslev, E. (2011). An aboriginal Australian genome reveals separate human dispersals into asia. *Science.* Doi:10.1126/science.1211177.

Ripple, W. J., & Van Valkenburgh, B. (2010). Linking Top-down Forces to the Pleistocene Megafaunal Extinctions. (Cover story). *Bioscience,*

60(7), 516-526.

Robins, S., & Mayer, R. E. (2000). The metaphor framing effect: metaphorical reasoning about text-based dilemmas. *Discourse Processes,* 30(1), 57-86.

Rockerfeller, J. (2009). The disappearance of food: The next global wild card? *The Futurist, 43*(3), 21.

Roszak, T. (1992). *The Voice of the Earth.* New York: Simon & Schuster.

Ryle, G. (1949). *The Concept of Mind.* New York: Harper & Row.

Saville, B. K., Gisbert, A., Kopp, J., & Telesco, C. (2010). Internet addiction and delay discounting in college students. *Psychological Record,* 60(2), 273-286.

Seligman, M. E. P. (1975). *On Depression, Development, and Death.* San Francisco: Freeman.

Shepard, P. (1982). *Nature and Madness.* Athens Georgia: University of Georgia Press.

Shepard, P. (1998). *Coming Home to the Pleistocene.* Washington, D.C.: Island Press.

Sjöblom, T. (2007). Spandrels, gazelles and flying buttresses: Religion as adaptation or as a by-product. *Journal of Cognition & Culture,* 7(3/4), 293-312.

Smith, C., Organ, D. W., & Near, J. P. (1983). Organizational citizenship behavior: Its nature and antecedents. *Journal of Applied Psychology,* 68(4), 653-663.

Steffen, W., Crutzen, P. J., & McNeill, J. R. (2007). The Anthropocene: Are humans now overwhelming the great forces of nature. *Ambio, 36,* 614-621.

Sundquist, K., Frank, G., & Sundquist, J. (2004). Urbanization and the incidence of psychosis and depression: Follow-up study of 4.4 million women and men in Sweden. *British Journal of Psychiatry, 184,* 293-298.

Taleb, N. N. (2007). *The Black Swan: The Impact of the Highly Improbable.* New York: Random House.

Teasdale, J. D. (1978). Effects of real and recalled success on learned helplessness and depression. *Journal of Abnormal Psychology,* 87(1), 155-164.

Thorens, G., Khazaal, Y., Billieux, J., Van der Linden, M., & Zullino, D. (2009). Swiss Psychiatrists' Beliefs and Attitudes About Internet Addiction. *Psychiatric Quarterly,* 80(2), 117-123.

Tiedens, L. Z., Unzueta, M. M., & Young, M. J. (2007). An unconscious desire for hierarchy? The motivated perception of dominance complementarity in task partners. *Journal of Personality and Social Psychology,* 93(3), 402-414.

Tooby, J, & Cosmides, L. (1992). The psychological foundations of culture. In J. H. Barkaw, L. Cosmides & J. Tooby (Eds.). *The Adapted Mind: Evolutionary Psychology and the Generation of Culture.* New York: Oxford University Press.

Tversky, A. & Kahneman, D. (1974). Judgment under uncertainty: Heuristics and biases. *Science, 185,* 1124-1131

Ulrich, R. S. (1984). View through a window may influence recovery from surgery. *Science, 224,* 420-421.

van den Eijnden, R. M., Meerkerk, G., Vermulst, A. A., Spijkerman, R., & Engels, R. E. (2008). Online communication, compulsive internet use, and psychosocial well-being among adolescents: A longitudinal study. *Developmental Psychology*, 44(3), 655-665.

Waller, J.E. (2004). Our ancestral shadow: Hate and human nature in evolutionary psychology. J*ournal of Hate Studies. 3,* 121-132.

Whitehead, A. N. (1929). *Process and Reality.* New York: The Free Press.

Wilkins, W. K., & Wakefield, J. (1996). Further issues in neurolinguistic preconditions. *Behavioral & Brain Sciences*, 19(4), 793.

Williams, E. (1977). Experimental comparisons of face-to-face and mediated communication: A review. *Psychological Bulletin,* 84(5), 963-976.

Wilson, E.O. (1984). *Biophilia.* Cambridge: Harvard University Press.

Wimberley, Fulkerson, and Morris, (2007), Predicting a moving target: Postscript for *The Rural Sociologist* on global-rural transition dates. *The Rural Sociologist, 28*, 18-22.

Winner, L. (1977). Autonomous Technology. Cambridge: MIT Press.

Zimbardo, P. G. (1973). The psychology of power and the pathology of imprisonment. In E. Aronson & R. Helmreich (Eds.), *Social Psychology.* New York: Van Nostrand.

Zimbardo, P. G. (1974). On "Obedience to authority.". *American Psychologist,* 29(7), 566-567.

Zualkernan, I. A., & Johnson, P. E. (1992). Metaphors of reasoning as problem-solving tools. *Metaphor & Symbolic Activity*, 7(3/4), 157.

Index

www.ingramcontent.com/pod-product-compliance
Lightning Source LLC
Chambersburg PA
CBHW031158270326
41931CB00006B/318